ISBN 978-1-334-64343-9
PIBN 10766488

1 MONTH OF
FREE
READING

at
www.ForgottenBooks.com

By purchasing this book you are eligible for one month membership to ForgottenBooks.com, giving you unlimited access to our entire collection of over 700,000 titles via our web site and mobile apps.

To claim your free month visit:

www.forgottenbooks.com/free766488

English
Français
Deutsche
Italiano
Español
Português

www.forgottenbooks.com

Mythology Photography **Fiction**
Fishing Christianity **Art** Cooking
Essays Buddhism Freemasonry
Medicine **Biology** Music **Ancient
Egypt** Evolution Carpentry Physics
Dance Geology **Mathematics** Fitness
Shakespeare **Folklore** Yoga Marketing
Confidence Immortality Biographies
Poetry **Psychology** Witchcraft
Electronics Chemistry History **Law**
Accounting **Philosophy** Anthropology
Alchemy Drama Quantum Mechanics
Atheism Sexual Health **Ancient History**
Entrepreneurship Languages Sport
Paleontology Needlework Islam
Metaphysics Investment Archaeology
Parenting Statistics Criminology
Motivational

THE AUKS

[ORDER: *Charadriiformes*. FAMILY: *Alcidæ*]

PRELIMINARY CLASSIFIED NOTES

[F. C. R. JOURDAIN.   F. B. KIRKMAN.   W. P. PYCRAFT.   A. L. THOMSON]

RAZORBILL [*Alca torda* Linnæus.   Razorbilled murre, marrot, auk, willock, razorbilled-auk.   French, *pingouin commun*; German, *Tord-Alk*; Italian, *gazza marina, tuffetto*].

1. Description.—The adult razorbill may be distinguished at once by the deep, grooved beak, marked with white lines, and the white line running from the beak to the eye. (Pl. 95.)   Length 17 in. [431·79 mm.].   The sexes are alike.   In summer the upper parts are black with a greenish gloss, save the throat and fore-neck, which are of a deep, velvety, brown hue.   The secondaries are tipped with white, and the lower throat and rest of the under parts are white, while a narrow, sharply defined line of white runs from the base of the beak to the eye, and a narrow, semicircular line of white, in a deep groove, crosses the beak, which otherwise is black, and is bounded behind by a raised fillet.   Two shallow grooves run across the beak in front of the white groove.   In winter the sides of the head, the throat, and fore-neck are white, the rest of the plumage is as in summer, but lacks the greenish gloss. The white line from the beak to the eye is only barely traceable, and the fillet round the base of the beak is wanting.   The iris is hazel, and the legs and feet are blackish or blackish brown.   The young in its first (protoptyle) down plumage has the head, neck, and under parts dull buffy white, the back dark brown, darkest on the rump. The second (mesoptyle) dress is coloured like that of the adult in summer, but rather browner.   The beak is relatively much smaller, and lacks the characteristic grooves and white markings, but the white line from the beak to the eye is well defined.   By the autumn the young have come to resemble the adult in winter dress, but may be distinguished by the beak which lacks the grooves, though the white line from the beak to the eye is fairly distinct.   [W. P. P.]

2. Distribution.—During the breeding season this species, like the guillemot, is found breeding on most of the rock-bound coasts and islets in our group, although in smaller numbers. As a rule, its breeding-places are also higher up on the cliffs than those of the kittiwake, below those of the puffin, and often overlapping the vertical range of the guillemot.   Outside the British Isles it also breeds in the Færoes and Iceland, and along the Norwegian coast from Stavanger Fjord northward to the North Cape (Collett), east of the Varanger Fjord in Russian Lapland (Pearson and Pleske), and perhaps also on the Murman coast, while, according to Buturlin, it is found in the Kola Peninsula and the White Sea to Onega Bay.   Southward it is found on the coasts of Brittany, as well as on Bornholm, Helgoland, Gotland, and other localities in the Baltic, even in the Gulf of Bothnia and along the coast of Finland.   On the west side of the Atlantic it is found from 73° N. on the west coast of Greenland along the coast of Labrador south to Newfoundland and the Bay of Fundy.   Its winter range extends from the North Sea and English Channel southward to the Bay of Biscay and the Straits of Gibraltar, while in the Mediterranean it has occurred as far east as the Adriatic Sea and Malta.   [F. C. R. J.]

3. **Migration.**—Resident, in that examples may be seen in territorial waters all the year round ; but our information regarding the species' movements is very incomplete.   From late in March till early in August the birds are at their breeding-places, but for the rest of the year they remain at sea.   In rough weather they become numerous close inshore, and their bodies are frequently thrown up on the beach ; sometimes they are seen in estuarine waters or even inland.   But the greater proportion of the birds are out at sea ; little information exists as to how far they go, and we can only conjecture whether British waters are visited by birds from farther north.   There is some evidence that our own birds remain near our coasts during the autumn and move southwards in December, after which it is principally young birds that are to be found (cf. Ussher and Warren, B. *of Ireland,* 1900, p. 357).   [A. L. T.]

4. **Nest and Eggs.**—No nest is made by the razorbill : the egg is laid on the bare rock, but if possible in some sheltered crevice, or, at any rate, on a much overhung ledge, so that it is often difficult to reach. (Pl. xxxix.)   Apparently in places where crevices in the rocks do not exist, it breeds in holes like the puffin (cf. H. J. Pearson, *Three Summers among the Birds of Russian Lapland,* p. 16), and more frequently under boulders, as at Scilly (see p. 15).   Only one egg is laid, less pyriform and more conical in shape than those of the guillemot, and showing less variety in colour and markings.   The ground-colour usually varies from white to yellowish, pale brownish or reddish, blotched, spotted, and smeared, often in the form of a

PLATE XXXIX

Photo by Riley Fortune

Razorbill's young and eggshell on an open ledge

Photo by Bentley Beetham

Razorbill with its eggs in a crevice, and a pair of guillemots below

Photo by F B Kirkman

Guillemots on the Pinnacles, Farne Islands

zone at the big end, with rich chocolate-brown or black. In some cases the markings are so numerous that they obscure the ground-colour altogether, while in rare cases a bluish green ground tint is observed. The interior of the shell looks greenish when held up to the light, while in guillemots' eggs the corresponding colour is yellowish white, except when the shell is coloured blue-green. (Pl. F.) Average size of 100 eggs, 2·95 × 1·86 in. [75· × 47·4 mm.]. The breeding season on our coasts begins a little earlier than that of the guillemot : probably the second to third week of May is the average time, but in the high north not till a month later. Although only one brood is normally reared during the season, a second and third egg is usually laid if the first is destroyed. Incubation is performed by both sexes, and, according to W. Evans, lasts 30 days ; while F. G. Paynter observed it for 25 days. [F. C. R. J.]

5. **Food.**—Chiefly small fish and crustaceans. The species drinks salt water. The young are fed by both parents on small fish. [F. B. K.]

GUILLEMOT [*Uria troile* (Linnæus). Scout, willock, willie, murre, tinkershere ; eligoog (S. Wales) ; longie (Shetlands) ; stronnag (Isle of Man). French, *guillemot à capuchon* ; German, *Schmalschnabel-Lumme* ; Italian, *uria*].

1. **Description.**—The guillemot is to be distinguished from its congeners by the relatively long slender bill, which is without grooves or markings of any kind. (Pl. 94.) Length 18 in. [457·19 mm.]. The sexes are alike. There is a distinct summer and winter plumage. In the summer dress the upper parts, including the head and neck, are of a dark slate-grey, but the fore-part of the head and throat and of the fore-neck have a decided tinge of smoke-brown. As the season advances the brown hue increases, the slate-grey assuming a brown tinge also. The secondaries are tipped with white, forming a bar across the wing. The lower part of the fore-neck and the rest of the under parts are white, but the uppermost flank-feathers are slate coloured. In winter the sides of the head, throat, and upper part of the fore-neck are white, but the white on the side of the head is interrupted by a dark triangular slate-coloured patch extending from the eye backwards above the ear-coverts. The young in its first (protoptyle) plumage has the crown and back of the neck dark brown relieved by yellowish white hair-like rami ; the back dark brown, the sides and front of the neck dull white with numerous distinct but narrow stripes of dull black. The breast and abdomen are white. The succeeding mesoptyle plumage (which has been mistaken for a true teleoptyle dress) is coloured like

that of the adult in summer. In the autumn the teleoptyle dress is assumed,
which resembles that of the winter dress of the adults, but the feathers of the throat
have faint dusky mottlings. [w. p. p.]

  2. **Distribution.**—In the British Isles this species is found breeding wherever
cliffs and steep rock faces afford sites for breeding colonies, both on the mainland
and on isolated stacks, and in some districts, such as the great range of chalk cliffs
from Speeton to Bempton in Yorkshire, and in various parts of the Welsh, Scotch,
and Irish coasts, enormous numbers may be found breeding close together. Outside the British Isles it is found in Iceland and the Færoes, on Helgoland, Bornholm,
Store-Carlsö, and Gotland, and along the Scandinavian coast up to the Varanger
Fjord, while it is said also to have bred in the E. Murman coast. Southward it
occurs on the northern and north-western coasts of France, and eggs have been
obtained on the Berlengas Isles, off the coast of Portugal. On the American side it
is found from New England and Nova Scotia north to lat. 80°, while an allied form
inhabits the North Pacific. The winter range of this bird extends to about lat. 30°
in the Atlantic, but it is scarce in the Western Mediterranean. [f. c. r. j.]

  3. **Migration.**—The remarks on the razorbill under this head apply equally
to this species. A guillemot " ringed " as a chick on the Aberdeenshire cliffs in
July 1910 was shot near Gothenburg on November 29 of the same year (*Aberdeen
University Bird-Migration Inquiry*). [a. l. t.]

  4. **Nest and Eggs.**—The guillemot makes no nest, but lays its single egg
on a bare shelf of rock, generally whitewashed with the droppings of the birds,
on a steep cliff face.   At the Farne Islands vast numbers nest on the top of the
Pinnacles, packed so tightly together that movement is impossible. The egg is
very large, and pyriform in shape, and the incubating bird pushes it between
her legs, with the pointed end projecting forwards, and sits usually with her back
to the sea. (Pl. xxxix.) In colour and markings these eggs probably vary more
than those of any other species, but can be classified under eight heads, according
to the ground-colour, which varies from pure white to creamy, yellow, or deep
blue-green. Sometimes eggs are met with entirely devoid of markings, others are
spotted or blotched, and in many cases an elaborate system of interlacing lines
covers the shell.   These markings also vary in colour from rich red to brown,
deep black, or some shade of dirty green, and in some cases they practically cover
the shell and conceal the ground-colour altogether. (Pl. G.) Average size of 122
eggs, 3·04 × 1·94 in. [77·37 × 49·45 mm.]. Incubation, which is carried on by both
sexes, lasts from 30 to 33 days, according to experiments with an incubator

(W. Evans), and Saxby confirms this, having found an egg with living young after 30 days' incubation ; while F. G. Paynter estimates it as 32 days. The first eggs are laid about 20th May on the Yorkshire coast, and are regularly collected till July. Only one brood is normally reared in the season, but if the first egg is destroyed, a second, and a third if necessary, are laid at intervals of about fourteen, or, according to E. W. Wade, nineteen days. [F. C. R. J.]

5. **Food.**—Chiefly small fishes, also crustaceans. The species drinks salt water. The young are fed by both parents on small fish. [F. B. K.]

GREAT-AUK [*Alca impennis* Linnæus. Penguin, garefowl, pinwing, northern auk. French, *grand pingouin*; German, *Riesen-Alk* ; Italian, *gran pingouino*].

1. **Description.**—The great-auk closely resembled the razorbill, from which,

THE GREAT-AUK.

however, it differed in its much greater size and the vestigial condition of the wings, which were useless for flight. The sexes were alike, and there were distinct "summer" and "winter" plumages. (See Fig. above.) Length 32 in. [933 mm.].

The summer dress differed from that of the razorbill only in that the white stripe in front of the eye was enlarged to form a large oval patch of white, while the white lines on the beak seem to have been rather less conspicuous.　In immature birds the feathers of the upper parts had pale margins, and the plumage generally was browner, while the white patch on the face was only faintly indicated.　The beak was smaller than in the adult, and lacked the grooves and ridges.　The winter dress was like that of the razorbill.　The young in down is said to have been of a dark grey colour. [W. P. P.]

2 and 3. **Distribution and Migration.**—Symington Grieve, in his monograph on " The Great Auk or Garefowl," defines the limits within which this extinct species is supposed to have occurred, as follows:—From Disco Island[1] along the west coast of Greenland and eastward to Cape Dan; from thence a line to the North Cape in Norway marks its supposed limits in the N. Atlantic; southward along the Norwegian coast and that of Sweden south to Malmö, including the Danish coasts; thence along the German, Dutch, Belgian, French, and Spanish coasts to Cape Finisterre, while a line from here to Cape Cod defines the southern limits of the species; along the coast of New England, the mouth of the St. Lawrence, and the Labrador coast to Kyucktabuck, and thence north possibly to Disco Island (?). Since then the southern limit on the American coast has been extended by the discovery of bones on the Carolina and Florida coasts.　The chief breeding-places within these limits appear to have been Eldey, Geirfuglasker, Fuglasker, and the Westmann Isles off the coast of Iceland; Sandöe in the Færoes, Papa Westray in the Orkneys, St. Kilda; and on the west side of the Atlantic, Funk Island, Penguin Islands, and "Aponars," off Newfoundland, Cape Breton, possibly the Bird Islands in the Gulf of St. Lawrence, and Cape Cod; and the Darrells or Graahs Isles on the east coast of Greenland.　Of these, only the Iceland, Fäeroe, Orkney, St. Kilda, and Funk Island stations are regarded by Blasius as certainly known. [F. C. R. J.]

4. **Nest and Eggs.**—The great-auk made no nest, but laid its single egg on shelving rocks or low-lying coasts within a short distance of the sea.　The breeding season is said to have been in June.　At some of its breeding-places large numbers appear to have bred close together, as guillemots do at the present day, and this seems to have contributed to their extermination, as expeditions were fitted out to kill off the birds for food during the breeding season.　Steenstrup states that when the egg or young were destroyed, no attempt was made to lay a

---

[1] More recent researches tend to prove that Disco Island should be deleted, as it is extremely doubtful whether the species has ever been obtained there.

second time. Average size of 29 eggs, 4·89 × 3·01 in. [124·3 × 76·4 mm.). In shape they resemble the eggs of the razorbill, but are of course much larger, and the ground-colour varies from whitish to stone-grey, yellowish, yellowish brown, and reddish yellow, while some eggs show faint traces of bluish green. In many cases exposure to light for long periods has resulted in the colours fading. The markings vary much in character. Some eggs are very sparingly marked, others exceedingly richly with spots, blotches, streaks, and smears of deep sepia-brown shading into black, which are sometimes concentrated at the big end. The presence of incubation spots in skins of both sexes shows that both male and female took part in incubation. [F. C. R. J.]

5. **Food.**—Little authentic information is available with regard to the food of this species, but there is small doubt that its diet consisted chiefly of fish. Amongst the species recorded are the lumpfish (*Cyclopterus lumpus*), the sea-scorpion (*Cottus scorpio*), and some species of the herring genus (*Clupea*). Blasius also suggests that crabs were eaten, as well as other lower forms of marine animals. [F. C. R. J.]

BLACK-GUILLEMOT [*Cepphus grylle* (Linnæus). Dovekie, Greenland dove, tystie, sea-pigeon. French, *guillemot à miroir blanc* ; German, *Gryll-Teist*]. [No Italian name is given, as it has not been recorded from that country. F. C. R. J.]

1. **Description.**—The black-guillemot is readily distinguished from its allies by the large patch of white in the wing, and the vermilion-red legs. The sexes are alike, and there is a distinct summer and winter dress. (Pl. 96.) Length 11 in. [279·40 mm.]. The adult male, in summer, is of a sooty black, with a green iridescent sheen ; and all the wing-coverts of the fore-arm, save only those of the marginal series, are white, forming a large and conspicuous white patch. The under wing-coverts and axillaries are also white. The eye is dark brown, the feet vermilion-red, the beak black, the inside of the mouth reddish orange. In the winter dress the crown, nape, and back of the neck are black with broad white fringes, giving the plumage a hoary appearance. The scapulars are black, broadly edged with white. The feathers of the lower back are black, broadly tipped with white, and on the rump the black is entirely obscured by white. The wings as in summer. There is a black patch in front of the eye. The under parts are pure white. This plumage is worn but for a very short period, and the transitional state presents an indescribable admixture of black and white, especially on the under parts. During the moult the

white wing-patch is cut into two by a black bar formed by the basal portion of the major coverts, which are exposed by loss of the white, overlying median coverts. Legs vermilion, as in summer. The juvenile plumage resembles that of the adult in winter. But the scapulars are black with a spot of white at the tip of each feather, giving a mottled appearance quite unlike that of the adult, while the white wing-patch is hardly traceable, being obscured by black, which colours the terminal portion of the feathers at this stage. Similarly, black tips to the feathers of the fore-neck, fore-breast, and flanks give a mottled appearance wanting in the adult. Inside mouth blush red, feet and toes deep brown. The first down (protoptyle) dress is of a uniform sooty brown hue. [w. p. p.]

2. **Distribution.**—In our Islands this species is confined to Scotland, Ireland, and the Isle of Man, and does not breed in England or Wales. In Scotland it is now absent as a breeding species from the east coast, except in Caithness, but is plentiful in the Orkneys and Shetlands, along the north and north-west coasts, and on the Hebrides, but becomes scarcer in the south-west. In Ireland, according to Ussher, it breeds locally in small numbers round the coast, but is much more frequent in the north and west of the country. Outside our limits it is known to breed in the Færoes, Iceland, along the Scandinavian coast up to the North Cape, and also eastward to the Varanger Fjord, in the Kola peninsula and the shores of the White Sea, on some of the Danish islands, and in the Baltic Sea along the southern coast of Sweden northward to Sundsvall at least, on Bornholm, its most southerly breeding-place, the Aland Isles, Karlö, and the coast of Finland in the Gulf of Bothnia. North of these limits it is replaced by Mandt's guillemot, but on the west side of the Atlantic it occurs from Massachusetts to southern Greenland, and breeds from Newfoundland to the Labrador coast as well as in Greenland. Its winter quarters are in the North Atlantic, and it rarely occurs south of the English Channel. [F. C. R. J.]

3. **Migration.**—Resident: the winter wanderings of this species appear to be of slight extent, as it is rarely met with in southern English waters. [A. L. T.]

4. **Nest and Eggs.**—No nest is made, and the two eggs are deposited on the bare rocks under boulders at the foot of cliffs, or in crevices of the rocks, and occasionally among ruins. As a rule, the site is low down and not far from the sea, but in the Shetlands it has been known to breed a hundred feet above the sea (*Zoologist*, 1891, p. 134), while Saunders says it has been known to nest a hundred yards inland. (Pl. XL.) The eggs are usually two in number, but three have occasionally been found together, and are whitish in ground-colour, sometimes

PLATE XL

Photo by Bentley Beetham

Black-guillemot's nest-hole and eggs

Photo by Bentley Beetham

Black-guillemot's nest-hole and young

with a tinge of bluish green, blotched and spotted with ashy grey and very dark blackish brown, which sometimes tends to form a zone. (Pl. F.) Average size of 52 eggs, 2·32 × 1·56 in. [58·9 × 39·8 mm.]. The breeding season is late, and in Scotland eggs are rarely laid before the end of May, and in Iceland from the beginning of June onwards. Incubation is carried on by both sexes in turn, and Hantzsch found the males sitting at night and the females by day in several cases. The incubation period is estimated by the same writer at three and a half weeks—24-25 days. One brood is reared during the season, but a second clutch of eggs is laid if the first is destroyed. [F. C. R. J.]

5. **Food.**—Chiefly crustacea and small fish. Also marine worms,· molluscs (Naumann, *Vögel Mitteleuropas*, xii. 239), and seaweed (E. Selous, *Bird Watcher in the Shetlands*, p. 203). The young are fed by both parents on small fish. [F. B. K.]

LITTLE-AUK [*Alle alle* (Linnæus); *Mergulus alle* (Linnæus). Rotge, rotchie; Iceland auk (Yorkshire); dovekie. French, *mergule nain*; German, *Krabbentaucher*; Italian, *gazza marina minore*].

1. **Description.**—The small size of this species and the short beak alone suffice to distinguish it from its congeners. (Pl. 96.) Length 7·5 in. [179 mm.]. The sexes are alike, and there is a distinct summer and winter dress. In the summer dress the head and neck to the fore-breast and the upper parts, are black, the back and wings have a greenish gloss, and are further relieved by white edgings to the hinder scapulars, and white tips to the secondaries. There is also a white spot over the eye. The uppermost flank-feathers have their upper margin slate coloured. The rest of the plumage is pure white. Iris hazel, legs and toes livid brown, the webs darker. In winter the throat and fore-neck are white, like the rest of the under parts. The juvenile plumage resembles that of the adult in summer, and this appears to be a mesoptyle dress, as in the guillemot and razorbill. The down (protoptyle) plumage is of a uniform sooty brownish black. Immature birds after the autumn moult are to be distinguished from adults in winter by the absence of the white spot over the eye. [W. P. P.]

2. **Distribution.**—To the British Isles this species is only a winter visitor; its breeding haunts lie in the high north, and include Grimsey, off the Iceland coast, and possibly also on the mainland (cf. *Ibis*, 1911, p. 8), Spitzbergen, the west side of Novaya Zemlya, Franz-Josef Land, Jan Mayen, Mevenklint, and in

Greenland from about lat. 68° to Foulke Fjord, lat. 78° 18'. The limits of its winter migrations extend on the west of the Atlantic to New Jersey, and on the east south to the Azores and Canaries exceptionally, although not recorded from the Straits of Gibraltar. It has, however, been occasionally met with in the Mediterranean. [F. C. R. J.]

3. **Migration.**—A winter visitor from the Arctic, mainly to our northern seas and coasts. The number of birds that reaches our area is very variable, noteworthy visitations taking place in some years. The winters 1841-2, 1894-5, and 1899-1900 were remarkable in this respect: less so were 1848, 1861-2, 1878-9, 1884-5, 1889-90, and 1900-1 (cf. Witherby and Ticehurst, *British Birds*, ii. p. 332; Ticehurst, *B. of Kent*, 1909, p. 533; Nelson, *B. of Yorks.*, 1907, p. 731; and Ussher and Warren, *B. of Ireland*, 1900, p. 368). November, December, and January are the chief months in which the species is found in British waters. [A. L. T.]

4. **Nest and Eggs.**—Does not breed in the British Isles. [F. C. R. J.]

5. **Food.**—The main food of this species consists of minute crustacea, and during the breeding season these are brought to the nesting-place by both parents in the distended cheek. Other plankton organisms are also devoured, and, according to Von Heuglin, Mollusca form part of its diet. Kolthoff notes that Amphipoda are eaten, and Collett found in the stomachs of Norwegian specimens Crustacea (*Euphausia inermis*) and traces of fish spawn. Among other genera of Crustacea which have been noted may be mentioned *Crangon, Palaemon, Mysis*, etc. Saunders states that during the autumn and winter months animal offal is also eaten. [F. C. R. J.]

PUFFIN [*Fratercula arctica* (Linnæus). Sea-parrot, coulter-neb, pope, bottlenose, Tammie Norie (Shetlands), pibbin (Isle of Man). French, *macareux moine;* German, *arktischer Lund*; Italian, *polcinella di mare*].

1. **Description.**—The puffin, when adult, may be recognised at once by the large size and brilliant colouring of the beak, and the bright orange-red legs. The sexes are alike. (Pl. 97.) Length 13 in. (330·19). There is no seasonal change of plumage. The sides of the face are of a pale smoke-brown colour, the crown, nape, a band round the neck, and the rest of the upper parts black; the uppermost flank-feathers are slate coloured, the rest of the under parts pure white. The upper jaw is bounded behind by a raised fillet, cream coloured, and finely pitted. This is succeeded by a triangular plate of pale blue-grey, the rest of the beak being red,

and marked by two deep grooves, while a similar groove, yellowish in colour, bounds the front of the triangular plate of blue-grey. The lower jaw is blue-grey at the base, bounded anteriorly by a raised margin of dull creamy white; at its tip is a conspicuous notch. On the upper eyelid is a triangular horny plate of blue-grey, and along the lower lid runs an oblong horny plate of the same hue. The rim of the eyelid is red, and at the gape is a rosette-like wattle of red. At the autumn moult the horny plates of the eyelids, the fillet at the base of the upper jaw, and the blue plates, are shed, thus materially reducing the size of the beak (see p. 51). The juvenile plumage differs from that of the adults in having a large dark patch in front of the eye, while the beak is much smaller, dagger-like, and lacks the brilliant coloration of that of the adults. The down (protoptyle) dress is of a uniform greyish brown, the breast white. This down is of great length, woolly in character, and probably represents both proto- and mesoptyle generations. [W. P. P.]

2. **Distribution.**—In the British Isles, the puffin, like the guillemot and razorbill, is chiefly confined to those parts of the coast where cliffs are found, so that it is absent from the low shores of Lincoln and East Anglia. It is not found in great numbers on the south coast of England, but on the other hand is met with in enormous colonies in the west of Scotland and many parts of the Welsh and Irish shores, also on the Scillies. Outside the British Isles it breeds in the Færoes, along the Norwegian coast, and in Bohuslan on that of Sweden, in Russian Lapland, on the west side of Novaya Zemlya according to Buturlin, as well as in Iceland, while an allied race, *F. arctica glacialis*, occurs in Spitzbergen, as well as in Greenland up to 70° N. Formerly many nested on Helgoland, while it is plentiful in some of the Channel Islands, and also breeds in Brittany, and it is possible that some also nest on the Berlengas. On the west side of the Atlantic its southern limit appears to be the Bay of Fundy. The winter quarters of this species extend to the New England States in N. America, and to the Azores, Canaries, and the Western Mediterranean on the east side. [F. C. R. J.]

3. **Migration.**—A summer visitor to our coasts, and uncommon in British waters during the winter months. Of the nature of its journeys almost nothing is known: as regards its return to its breeding-places, the late Professor Newton remarked, "Foul weather or fair, heat or cold, the puffins repair to some of their stations punctually on a given day as if their movements were regulated by clock-work" (*Dictionary of Birds*, 1896, p. 558). [A. L. T.]

4. **Nest and Eggs.**—Most nests of this species are to be found in burrows two or three feet long in the turf near the top of cliffs, but occasionally eggs may

be found in natural crevices or under stones.   Sometimes the single egg is laid on
the bare earth, but there is generally some loose grass and a feather or two.   According to Naumann, both sexes share in building.   In very crowded colonies eggs
have been found lying exposed on the bare rock.   (Pl. XLI.)   In colour most eggs
are a dull dirty white, rough, and quite without gloss, faintly blotched, spotted or
zoned with some shade of pale brown or violet, while eggs with darker and more
decided brown markings have been known to occur occasionally.   The eggs after
incubation are often deeply stained by the peaty soil.   (Pl. F.)   Average size
of 30 eggs, 2·39 × 1·67 in. [60·8 × 42·5 mm.].   The puffin returns to its breeding
haunts with great regularity, and begins to nest early in May, laying in the second
or third week of that month.   If the first egg is taken, another is laid about fourteen
days afterwards, but apparently not a third time.   Incubation is performed by
both sexes, and lasts, according to Messrs. F. G. Paynter and W. Evans, for 36
days.   Only one brood is normally reared in the season.   [F. C. R. J.]

   5. Food.—Chiefly fish and crustaceans.   The young are fed by both
parents on small fish.   [F. B. K.]

   The following is described in the supplementary chapter on " Rare Birds " : [1]—

   Brünnich's   guillemot,   *Uria   lomvia*   (Linnæus),   [*Uria   bruennichi*
   (E. Sabine)].

---

[1] The great-auk should more properly be placed under this head, but the interest attaching
to the species justifies the more complete treatment it receives by including it in the main body
of the book.

PLATE XLI

Site of puffin's nest under boulders

Same with boulders removed to show the nest and egg

Puffins' breeding-place, Annet, Scilly

THE RAZORBILL AND GUILLEMOT

[F. B. KIRKMAN]

In the older classifications the Auks were placed next to the Divers (*Colymbidæ*). A closer study of their structure has revealed the fact that, like the Gulls and Terns (*Laridæ*), they are a highly specialised form of the Plovers (*Limicolæ*). They are descended from some primitive plover that left the shore for the sea, the change from one habitat to the other being accompanied by a change in structure, which, by a process of convergent evolution, has led to their present resemblance to the divers, more particularly in respect to modifications tending to increase their powers of movement in the water, such as the lengthening of the body and the shifting back of the legs to its posterior end.[1]

The family is confined to the northern hemisphere. Of some twenty-five species, the names of seven find a place on the list of British birds. One of these, the great-auk, is extinct; a second, Brünnich's guillemot, is a very rare occasional visitor; and a third, the little-auk, a fairly common winter visitor. The remaining four species, the black-guillemot, puffin, and those which form the subject of this chapter, are to be found on British waters all the year round. All four belong to different genera, but the frequent association of the razorbill and common-guillemot makes it convenient to treat both together.

The two species are not difficult to distinguish, even apart from the marked difference in the shape of their bills.[2] The razorbill is a handsome bird, its rich black upper parts gaining an added effect in

---

[1] H. Gadow in Bronn's *Thier-reichs*, vol. vi.; *Vögel*, II. Systematischer Theil, p. 206; F. E. Beddard, *Structure and Classification of Birds*, p. 359; W. P. Pycraft, *History of Birds*, pp. 55, 450.

[2] What there is in the shape of the bill of the razorbill to justify its name I am unable to understand.

the breeding season from the clear-cut thin white lines round the beak, and from the beak to the eye. Its figure, if a trifle stout, is compact and well defined. The guillemot lacks finish, especially about the head and beak, owing to the unrelieved uniformity of the coloration of its upper parts. The ringed, bridled, or spectacled variety, which has a white ring round the eye and a white streak behind it, makes a better appearance, but these additions do not compensate for the want of a clear dividing-line between the head and the beak. The bridle marking has yet to be explained. The individuals which have it mix and breed freely with the ordinary type, but remain in a minority, which appears to vary, however, in different localities. The proportion of the variant to the typical form is given as one to fifty in Ireland, and one to five in the Western Isles.[1] At the Flamborough Cliffs (Bempton, May 1911) I saw scarcely any of the former, not more than three among some hundreds of birds.

Between the breeding seasons both species are to be found, usually in comparatively small parties, off our coasts or far out to sea.[2] They pass their days and nights on the water. Towards the beginning of February, or sometimes later, they draw near their breeding-haunts, which they appear to frequent somewhat irregularly till April, when they begin to occupy their nesting-sites in considerable numbers;[3] but laying does not begin till May.

Razorbills and guillemots are generally to be found on the same cliffs, but their relative numbers vary to a greater or less degree according to the nature of the nesting-sites provided. As is well known, guillemots prefer open ledges, or the tops of stacks, though they may also be seen in crevices or holes in the face of the cliff. Razorbills prefer the crevices, or a ledge which is overhung. On the Flamborough cliffs (Bempton) I have seen both

[1] Ussher, *Birds of Ireland*, p. 361; Harvie-Brown, *Outer Hebrides*, p. 161.
[2] The extent of their gregariousness between the breeding season is not very clear from the evidence. The same applies to the movements of the two species previous to their occupation of their breeding-haunts in April.
[3] Ussher, *Birds of Ireland*, pp. 356, 361; Nelson, *Birds of Yorks.*, ii. 711; *Annals of Scottish Natural History*, 1904, p. 22.

species on the same ledge, the guillemots on the half which was open, and the razorbills on the other half, which was overhung in such a way as to permit the birds to occupy it only in a sitting position. On these cliffs the former species is in a large majority. The opposite is the case at the Scilly Isles, "where precipitous rock faces are few in number, whilst rocks covered with huge boulders and intersected in all directions with nooks and crannies abound. Here, in the latter, razorbills are swarming, and are indeed often quite as numerous as the puffins. There are many low islets which only rise a few feet above the sea, and are scattered all over with boulders of various sizes, amongst which razorbills and puffins are about equally distributed, but where not one guillemot will be found."[1] On the Farnes, on the other hand, the guillemots are much the more numerous, for here they have not only suitable ledges but the flat top of the stacks known as the Pinnacles. The photograph of these on Pl. xxxix. and Mr. Seaby's drawing (Pl. 94) will give a good idea of the pressure upon the available space. To realise, however, the enormous numbers in which these birds, together with puffins and other cliff-frequenting species, congregate in the breeding season, one must go to larger colonies, like those in the Western Isles, where the birds crowd the sea and the cliffs, and blacken the air in their passage from one to the other. At N. Ronay, in the Outer Hebrides, the huge number of breeding guillemots is said to be augmented by the visitations of hundreds of thousands of non-breeding birds.[2] These include, no doubt, a certain proportion of immature individuals. In the case of razorbills, Mr. Ussher states that a certain number of the birds of the previous year make their appearance at the Irish breeding-haunts, but only in small numbers.[3] According to Naumann, the non-breeding birds may be seen in pairs, either sitting among the others, or in groups apart, the groups being mostly composed of the young of the previous year. He adds that these pairs perform just the same love actions,

[1] C. J. King *in litt.*
[2] Harvie-Brown and Buckley, *Fauna of the Outer Hebrides*, p. 161.
[3] *Birds of Ireland*, p. 356.

billing, bowing, etc., as those that hatch eggs.[1] The matter is well worth further attention.

There can be little or no doubt that the birds pair for life and return to the same ledges each year. Whether one sex returns before the other is not recorded, and there are, unfortunately, no published details of the struggle for nesting-sites that must take place at this period wherever there is overcrowding.

On one occasion I witnessed what was evidently a battle for the ownership of a crevice between two razorbills. It took the form of a tug-of-war, both birds crouching with their beaks apparently inter-locked. As the strain of the tugging fell chiefly upon the neck, the heavy bodies remained much where they were. After about twenty minutes of this ineffective struggle, both flew off together, alighted on the water, and there continued to battle, finally diving.

Owing to the similarity of the sexes, it was impossible to say whether this contest was between males or females, or between a male and a female. In studying contests for sites, the first important thing to ascertain, from the scientific point of view, is whether or not the males return to the breeding-haunts before the females. If they do, it may follow that conquest of a nesting-site carries with it possession of the female accustomed to lay there. The female, on her arrival, may find a new mate awaiting her, instead of the old. If she accepts him, as a matter of course it is clear that the theory of sexual selection, as far as these species are concerned, falls to the ground.[2] Whether such be the case or not, close observation of these and other cliff-birds at the beginning of the season would give very important results. Hitherto, unfortunately, their eggs have been regarded as of much more importance than the birds themselves. Hence our ignorance of their habits, though they are exceedingly easy to watch.

Fighting takes place at any time during the breeding season, and

---

[1] *Vögel Mitteleuropas*, xii. 221.
[2] For observations on the same point in respect to the Warblers, see H. E. Howard, *British Warblers*.

Plate 94

Guillemots on a corner of the Pinnacles, Farne Isles

By A. W. Seaby

seems to arise, in part at least, from one bird's objection to the presence of another near the particular spot which it regards as its own. It is difficult otherwise to explain the menaces and thrusts exchanged between adjacent incubating birds, or the pecks that a guillemot returning to its ledge often receives from those near which it alights, for no apparent reason except that before reaching its own particular spot it trespasses on the property of others.

Guillemots appear to be more combative than razorbills, no doubt because the fact that they breed in numbers on the same ledge or stack brings them much closer together, and consequently increases the occasions for strife. But the thrusts they make at one another are sometimes merely a matter of form, and at others mere play. I have seen a number thus playing upon the water. Two would point beaks at and swim round each other much like a couple of fencers seeking for an opening, but they went no further, and were evidently in high spirits. Just the same play may sometimes be seen when a pair greet each other.[1] More serious are the disputes that often follow the arrival on the ledge of a guillemot with a fish in its beak. On its passage to its mate or young it has to run the risk of having its prey snapped away by another bird, who in turn may suffer the same loss. Thus the fish may pass from bird to bird before it is finally engulfed. "Or it may be tugged at for a long time by two birds that have a firm hold of the head and tail part respectively, and pull it backwards and forwards, not infrequently across the neck of a third bird standing between them."[2]

As far as my observation goes, the guillemots' method of fighting is the same as the gannets'. Though they make passes with the beak closed, they do not strike with it closed, but either give a sharp nip if they can take their adversary unawares, or else seize him by the beak and tug. I have seen one bird with the beak of another almost inside its own. In this position they remained silently struggling for

[1] E. Selous, *Bird Watcher in the Shetlands*, p. 163.
[2] E. Selous, *Bird Watching*, p. 188.

two or three minutes, when, after a final tussle, the aggressor
decamped. The description given above of a battle between razorbills makes it appear that their method is the same.

Like other birds, guillemots and razorbills express the rebirth of
love in the spring by a variety of actions. These are usually called
courtship actions, but, as noted in another chapter,[1] the term is,
strictly speaking, inexact, for such actions occur in the case of birds
that live in pairs throughout the year, and continue after mating has
taken place. They are simply the outward and visible sign of the
birds' feelings, and may be witnessed at all times during the breeding
season. The display of the cock sparrow can, for instance, be seen
any day from January to July—that is, during nearly half the year.

One of the most usual expressions of mutual love among both
guillemots and razorbills is common to many birds. This is the
familiar "billing," which takes the form of touching, rubbing, or
interlocking bills, and corresponds to the human kiss. The movements of the head and neck that accompany it are sometimes
performed in the air without actual contact of the bill. Closely
connected with "billing," and legitimately included under the term,
are the caresses which take the form of nibbling with the tip of the
bill at the plumage of the beloved. I have more than once seen a
guillemot, seated in an attitude of slippered ease, on the ledge with its
head bent over its back, its beak pointing blissfully upward, and a
wide expanse of neck and throat presented to its mate, who
affectionately nibbled at it, to be afterwards caressed in turn. Sometimes a bird passes its bill over the head and neck of its mates
without nibbling. This, at least, certainly applies to razorbills,
probably also to guillemots. These love caresses are by no means
always gentle; indeed, one might well mistake the pecks then given
for an assault, did they not gradually develop into an exchange of
gentler pecks and nibblings. At times, of course, there are serious
domestic disputes, followed, however, by a happy reconciliation. A

[1] Vol. ii. 65, 66.

third form of caress is the twining by one bird of its neck round the neck of its mate, which I have noted in the case of guillemots, and also of gannets.   Here the avian neck plays the part of the human arm.

The most remarkable of these love actions was one which I witnessed in the case of guillemots only.   In its complete form it consisted in running the head down the breast till the tip of the beak nearly touched the ground between the feet, then in raising it well up and jerking it about quickly from side to side.   Sometimes the head was moved down a short way only, then raised and jerked.   Sometimes, again, the head might be jerked without any preliminary bowing action ; and sometimes the latter alone occurred, or was followed by some caress, such as the feather nibbling.   On one occasion I saw one bird keep its head between its feet while the other delicately nibbled at its crown.   These actions were not confined to one sex.   Both might be seen bending their heads to the ground side by side.   But the presence of both was not essential.   One would occasionally perform by itself.   The shorter, unfinished bows, curious forward movements of the head, were made not only by pairs but by a number at a time.   In the latter case it appeared to have rather a social than a sexual significance.

The downward inclination of the head to the feet is the most interesting feature in this display, on account of its close resemblance to the movement performed by guillemots when pushing their eggs beneath their breasts with the bill, before settling down to incubate. I have indeed seen guillemots, before they had eggs, shuffle along the ledge as if seeking a spot to lay, and then bring the beak down between the feet in the way described.   In these cases the action may have been actually anticipative of egg-laying, and have had no sexual meaning.   But circumstantial evidence left no room to doubt that in the majority of cases it was a love-display.   The resemblance between the two acts probably extends to the curious little nibbling movements that the bird often makes with the bill during the space of time in its display when the tip of its beak is held just above the

ground. As no guillemots had started to lay at the time I made the
preceding observations (May 9-13, 1911), I was unable to verify this
detail, but the assertion of probability seems to me more than justified
by the fact that I saw a razorbill, which species is also in the habit
of pushing its egg under its breast, perform the identical nibbling
motions before it actually put its bill on the farther side of the egg
in order to roll it into place over the side of its foot.

If this use of the same act for more than one purpose were an
isolated fact, it might pass as a mere coincidence. But it is not.
Birds, being creatures with limited means of expression, naturally
tend to make a familiar act do duty in different ways. The robin's
song, for example, expresses love in spring, and in winter is a battle
note or a sign of well-being. Further, the act may be primary in one
of its uses and derived in the other. For example, the bowing display
of the guillemot has all the appearance of being derived from the
more primitive act of pushing the egg into place. The jerking of the
head in the air that follows the bowing action may be explained in
the same way as a development from the nibbling caress, for, as we
have seen, the bird often makes nibbling movements without actual
contact. In the chapter on Gulls, similar examples will be given.
The subject cannot be pursued further here. It leads directly to the
very large question of the evolution of animal behaviour as distinct
from animal structure. It is one to which field ornithology could
make most important contributions.

The razorbill, besides the billing and the nibbling caress already
described, also jerks its head about. I have not seen it indulge in the
guillemots' curious bowing movements; indeed, it has less room for
display of any sort, owing to the less open nature of its nesting-place.
But it has seemed to me to be more vocal when in its love transports,
its chief effort being a harsh continuous croaking note. Generally
speaking, however, the guillemot is a much noisier bird, its note being
present to the ear almost continuously during the height of the breeding season. It is generally described as "*murr*," and as such has

given one of its names to the species. The description is very in-adequate, and overlooks both the fact that the bird utters more than one distinct sound and has a wide range of intonation.[1] Its most im-portant note bears for me a strong resemblance to the long-drawn meditative croon of the common barn-fowl. To Mr. E. Selous, on the other hand, "it resolves itself into a sort of *jodel*, long continued, and having a vibratory roll in it. It begins usually with one or two shorter notes, which have much the syllabic value of 'hărāh, hărāh'—first *ă* as in 'hat,' with the accent on the last syllable, as in 'hurrah.' Very commonly the outcry ends here, but otherwise the final 'rah' is prolonged into the sound I speak of, which continues rising and falling—which is why I call it a *jodel*—for a longer or shorter time, the volume of sound being increased, sometimes, to a wonderful extent. It ends usually as it began, with a few short rough notes. . . . "[2] The note not only strikes different ears in a different way, but varies very considerably for the same ear. I find in my note-books three attempts to syllable it, as follows :—

> "*Querr, quow, quow, hr hr hr hr.*"
> "*Krrowww, how, how how, how, how.*"
> "*Koo-werrrr-koo, koo, koo, koo.*"

A note that sometimes enters into it is a duck-like sound, syllabled by Mr. Selous as "*ik, ik, ik,*" and in my note-books by repetitions of either "*hec*" or "*uk*" or "*wik.*" This note may be used by itself. The species probably utters other distinct sounds. To tabulate its language adequately requires a phonetically trained ear and a phonetic notation. The present attempts to represent, in the traditional or nomic spelling, the sounds uttered either by this or any other species are practically useless for purposes of exact comparison.

Both guillemots and razorbills lay one egg only, unless robbed, when they will lay a second, and, if this is taken, even a third. No nest is built, the egg being deposited on the bare rock ; but guillemots

---

[1] By which is here meant the rising and falling of the voice in pitch.
[2] *Bird Watcher in the Shetlands*, pp. 113, 172, 187, etc.

have been seen to bring what might serve for nest material (feathers, grass, roots) to their mates, and lay it on the rock before them, where it is sometimes pulled about by both.[1]   This may be mere play, but it has at least some appearance of being the survival of a lapsed nest-building instinct, granted that the primitive plover from which the Auks are descended had acquired the instinct, which, after all, is an assumption.   If it had the instinct, there at once arises a difficulty in explaining why it should have lapsed.   That there is nothing in the nature of the nesting sites occupied by guillemots and razorbills to prohibit the construction of a nest is a fact to which the kittiwake, gannet, and puffin bear witness, particularly the two first named, which erect particularly solid structures on ledges often exactly similar to those which a guillemot might choose.   The kittiwake, let us note, like other gulls, is also descended from plover stock.   Possibly the Gulls branched off from stock in which the nest-building instinct had become well established, and the Auks did not.   The nest-building instinct of the puffin may have been subsequently acquired.

The fact that razorbills and guillemots lay one egg can be explained by the large size of the egg and by the difficulty of incubating more than one on bare rock-ledges, where the eggs would tend to roll apart in the absence of cup-shaped depressions such as are provided by a nest or a scrape in the soil.   Yet the gannet also lays one egg—in a nest!

The extraordinary variation in the coloration of guillemots' eggs has often been remarked, and provides one of the excuses for collecting them in large numbers, though the collectors remain, for the most part, singularly reticent about the scientific results of their labours. The variation extends not only to the colour and shape of the markings but also to the ground-colour, which may be white, blue, green, brown, yellow, buff, or pink.   (Pl. G.)   The variation in the razorbills' eggs are less marked, not more so than in those of many other species.

The variation, in the case of guillemots' eggs, has been explained

[1] E. Selous, *Bird Watching*, pp. 104-5 ; *Bird Watcher in the Shetlands*, p. 100.

by the supposition that it helps each bird to find its own among the many that may be crowded upon a ledge or stack. In support of this, Yarrell states the fact that birds marked with splashes of paint were found in their accustomed places day after day. The fact, however, does not necessarily prove his contention, for there can be little doubt that the birds return again and again to their accustomed places before their eggs are laid. No doubt each bird does recognise its own egg as long as it remains fairly clean, but, as Saxby pointed out long ago, the eggs tend in a short time to become so coated with filth, especially in wet weather, that they cease to show any noticeable variation. When they are in this condition it is probable that each bird would recognise the accustomed place much more readily than the egg. In any case, there is no clear evidence that the variation has been evolved to facilitate recognition. It may more plausibly be explained by the supposition that here natural selection has ceased to act. The ancestral plover, from which the Auks are descended, no doubt laid its eggs on the ground, and those survived whose eggs assimilated most closely in coloration to the site habitually chosen. Hence, as in the case of the Waders generally, the emergence of eggs of a *relatively* uniform coloration. But from the moment the ancestor of the Auks started to lay on open rock-ledges, chalk, red sandstone, limestone, and others, the coloration of its eggs would, owing to the change in the character of the site, cease to have protective value. The guillemot appears to have met the difficulty by systematic sharing by the sexes in incubation, so that when one of a pair was off the other was on, or usually so. This made the coloration of the egg a matter of little importance, and variation, previously checked by natural selection, had free play.

The pear shape of the guillemot's eggs, and the somewhat more conical shape of the razorbill's, serves, it has been suggested, to prevent them from rolling off the ledge into the sea by causing them to spin round when disturbed. In fact, the shape of the eggs does not prevent them falling off the ledges in cascades when a colony is

suddenly startled by the report of a gun or otherwise, for each egg
rests between and partly, if not wholly, on the bird's webbed feet.
The sight of such sudden downfalls was, and possibly still is in certain
places, "one of the amusements of the gaping tourist," to quote
Alfred Newton. When not frightened off its egg, the bird is careful
to get its feet away from it before taking flight. When thus left, a
pear-shaped egg would, of course, if the rock where it stood were not
level, be much less likely to roll than a round one, but it would be
rash to assume that this explains the pear shape, for it is almost
certainly that of the guillemot's ground-laying ancestor. Indeed,
the similarity of the eggs of Auks and existing Plovers is regarded
as one of the proofs of their descent from a common stock. The
primary use of the pyriform shape may be seen by examining the
arrangement in the nest of the four eggs of the green plover or lap-
wing (*Vanellus vulgaris*). They lie with their points meeting, and so
occupy the minimum of space.

Both guillemots and razorbills appear to sit with the egg not
across the webbed feet, but lengthwise between them. The guillemot
incubates either in an erect position or lying flat, the former, no
doubt, originating as a space-saving device on crowded ledges. It
usually sits with its breast facing the cliff. This is especially the
case on ledges sloping seaward. On less sloping cliffs its has been
seen to sit sideways, and on large roomy flat surfaces with the breast
to the sea.[1] The normal position of the incubating razorbill is the
recumbent. The nature of its nesting-place would in most cases
render any other impossible. Both sexes, in the case of both species,
share in incubation, and have brood-spots for the purpose. The males,
in addition, feed the hens when incubating.[2] Naumann denies this
in the case of the guillemot.[3] Personally I have watched incubating

---

[1] C. J. Patten, *Aquatic Birds*, p. 485.
[2] For the guillemot: Harvie-Brown and Buckley, *Fauna of Outer Hebrides*, p. 161 (Finlayson);
E. Selous, *Bird Watching*, p. 187. For the razorbill: Patten, *Aquatic Birds*, p. 470; H. Saunders,
Manual of British Birds, p. 696.
[3] *Vögel Mitteleuropas*, xii. 212.

Plate 95

Razorbills

By A. W. Seaby

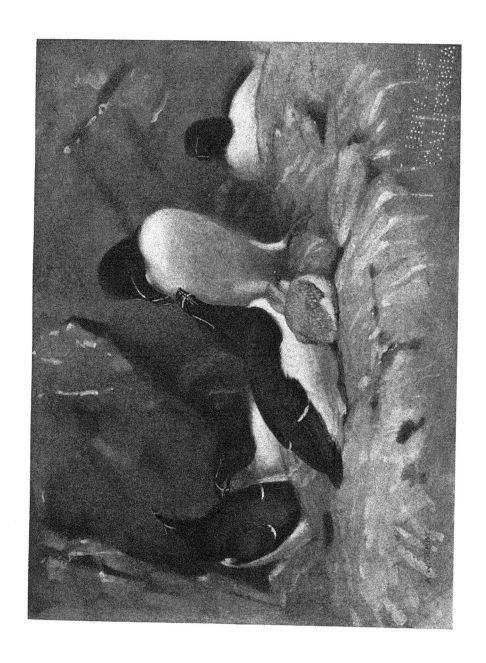

razorbills for some hours without seeing the sitting bird once fed.[1]
Perhaps the practice of individual birds varies.

The young are hatched after about 25-30 days' incubation. It
has been noted by Mr. Pycraft that, in the case of both, the second
or "mesoptyle" generation of prepennæ (down feathers replaced
later by contour feathers[2] is not only much less degenerate in structure than the first or "protoptyle" generation, but in coloration
exactly resembles the summer dress of the adults, the resemblance
extending, in the case of the razorbill, even to the white stripe in
front of the eye. This fact supports Mr. Pycraft's contention that
the "mesoptyle" down is a degenerate ancestral plumage, though
less degenerate than is usually the case.

The young remain on the ledges from three to four weeks, where
they are fed assiduously by both parents on small fish. The guillemot
brings one at a time, holding it lengthwise in the beak, with the tail
projecting about an inch and hanging over to one side. The razorbill
holds its prey across the bill; it is, therefore, able to bring more than
one fish, and does so—half a dozen at a time, the heads and tails
forming a fringe on either side. This marked difference between
the habits of the two species will be dealt with further in the chapter
on the puffin, which carries its fish in the same way as the razorbill.[3]

In pursuing their prey under the water both species use their
wings. They have the appearance of flying under the water, the legs
trailing behind, and used only—though about this I am not certain—
to help in turning. The resemblance of flight under the water to that
in the air is only superficial. The difference is thus stated by J. B.
Pettigrew: "In aerial flight the most effective stroke is delivered
downwards and *forwards* by the under, concave or biting surface of the
wing, which is turned in this direction; the less effective stroke being
delivered in an upward and forward direction by the upper, convex or

[1] The same is the experience of a writer in the *Annals of Scottish Natural History*, 1904, 22.
[2] Vol. i. p. 15.
[3] The curious fact, hitherto unnoticed, that the razorbill, like the puffin, sheds part of its
bill in winter, will be treated in the same chapter (p. 52).

non-biting surface of the wing. In sub-aquatic flight, on the contrary, the most effective stroke is delivered *downwards* and *backwards*, the least effective one upwards and forwards." In aerial flight the line of the body is inclined slightly upwards, in the other slightly downwards. The difference is explained by the fact that, while forward movement is the object in both cases, the sub-aquatic flier, being lighter than water, has to keep itself from rising, while the aerial flier, being heavier than the air, has to keep itself from falling.[1]

The fish, when brought by guillemots to the ledge, is either dropped there for the young to pick up, or is snatched by the latter before it touches ground. What is the procedure in the case of the razorbill I have been unable to find out.

Both species drink salt water. The same fact has been noted in the case of the sandgrouse.

The following are two little domestic scenes described from life by the pen of Mr. E. Selous, who has recorded in greater detail than any one the life of guillemots on the ledges during the time that the young are hatched.

The first illustrates the act of feeding the chick :—"The chick, when a very substantial fish is brought in for him, is asleep under his mother's wing, and both parents seem averse to disturbing him. The one with the fish seems quite embarrassed. He approaches, stands still, looks at his partner as if for advice, shuffles about, turns this way and that, and several times, bending his head, gives a choked and muffled *jodel*, for his mouth is almost too full to speak. Still the chick sleeps on, and still the parents seem to doubt the advisability of waking him. At length, however, they admit it to be necessary. The father shuffles up into his usual position, the mother rises by slow and reluctant stages, as though apologetically, and finally stirs the chick several times with her bill till at last he rouses. Then, in a moment, he busks up, and, seizing the large fish, swallows it in a whole-hearted gulp." [2]

[1] *Animal Locomotion*, p. 92.   [2] *The Bird Watcher in the Shetlands.*

The term "shuffle," it may be remarked parenthetically, is quite appropriately applied to the guillemot's mode of progression on land— and equally to the razorbill's. Both walk, and also stand, on the metatarsus as well as the toes, that is, on the whole sole of the foot, instead of, as in the case of most birds, on the toes only; or, in less exact language, the guillemot and razorbill walk on the foot and so-called "shank"—the two together constituting what corresponds to the mammalian foot, from the lower set of ankle-bones to the toes (Vol. I. p. 14, Fig. 1). Occasionally both species hop, but still alight on the metatarsus and toes. They have been seen sometimes to raise themselves up on the toes, and both walk and run in this attitude.[1] According to Dr. Saxby, who was a good observer, the young guillemots run and stand upon the toes; it is only as they grow older that they rest on the metatarsi as well.[2] This is a most interesting fact, and tends to show the adult bird's mode of progression is of comparatively recent date, but as to how it arose I can form no idea.

The second scene illustrates the affection shown by the parents for the young, which takes forms similar to the caresses of one sex by the other already described:—"After a while it (the young guillemot) comes out (from under its mother's wing), and the mother, as she stands by it, just stirs or nibbles the feathers of its face with the end of her bill—an action which has all the spirit of wiping a child's face or nose. The father now walks up, stops in front of the chick, bends down its head and *jodels*. Then it lifts it up and *jodels* more loudly; then, stopping again, preens the chick's head and face a little with the point of its bill, and nibbles at it affectionately." The chick then makes a little excursion along the ledge by itself, and then toddles back to its admiring parents, to be nibbled at and bowed and jodelled over once more.[3]

How young guillemots and razorbills quit the cliffs for the sea is a question about which there is still some difference of opinion. When

[1] E. Selous, *Bird Watching*, p. 195.
[2] *Birds of Shetland*, p. 248.
[3] *Bird Watcher in the Shetlands*, p. 178.

the exodus takes place, they are not fully fledged, and have little strength in their wings, not enough to permit them to fly off to the sea. Their usual method, according to the best available evidence, is to flutter and slide and fall down the side of the cliff to the waters below, an appalling journey for so inexperienced and puny a creature. That they should hesitate is not surprising. Their parents, however, are strong-minded enough to jostle them along.[1] Razorbills have been seen jostling their young all the way down to the sea, from foothold to foothold, so that the unhappy little birds went "rolling, and tumbling and falling sometimes down steep cliffs."[2] Professor C. J. Patten corroborates this: "The young are apparently roughly treated, being jostled and pushed off their ledges; yet their fall is so broken as they tumble and scramble down the face of the cliff that they generally reach the water in safety."[3] The evidence for the statement that guillemots and razorbills will entice their young on to their backs or seize them by the neck or wing and so carry them down by force is contradictory, and rests largely on the observation of fishermen, which is notoriously inexact. But it cannot be merely disregarded. It may still be shown that the recalcitrant young are treated in one or other of the above ways.

Perhaps the most noteworthy feature in the exodus of the young is that it occurs before they are fledged. If they were in possession of wings, like, for instance, the young gannet, the descent would present no dangers, and several lives would be saved. It is difficult to see what the species gain by this early flight, for the young are as much exposed to enemies, the larger gulls for instance, on the sea as on the cliffs, indeed more so. A possible explanation takes us once more back to the ancestral plover. If it resembled the present ground-nesting plovers, it would certainly have been in the habit of leading its young from their exposed birthplace long before they

[1] For the guillemot see C. J. Patten, *Aquatic Birds*, p.486; Naumann, *Vögel Mitteleuropas*, xii. 222; *Annals of Scottish Natural History*, 1905, 146. For the razorbill the evidence is more precise: C. J. Patten, *Aquatic Birds*, p. 470; *Zoologist*, 1871, 2427, quoted in Yarrell, iv. 55; *Irish Naturalist*, 1899, 132.

[2] *Irish Naturalist*, 1899, 132 (E. M'Carron).   [3] *Loc. cit.*

were fledged.   The. instinct to do so may still survive in the razor-bills and guillemots.

Their first experience of the water appears to be anything but agreeable to the young.   Professor Patten writes of the young razor-bill: "It is amusing to watch the bewildered expression of the youngsters when they receive their first ducking in the briny deep. I have heard them crying in piteous accents for their fond parents, who, out of their sight in the depths beneath, were diligently pursuing fish to feed them."[1] The paucity of the evidence, and its contradictory nature, makes it impossible at present to say with any certainty whether the young dive and feed themselves instinctively, or whether they have to learn.   Naumann states explicitly that young razorbills begin to dive for food in the water, though a few minutes before accustomed to receive their food on the rock.   He adds that the parents do not feed them.[2] This, however, is not the view of others.

The object of the parents appears to be to get their offspring away to sea ; and as to the manner in which the razorbill does this we have the following interesting first-hand evidence : "The old bird wants to get the young one off to sea.   The young one, apparently, does not understand this, and merely swims about.   The old bird seems excited, swims round it and right off before it a few yards, then returns and dives a few times round about it.   At last it commences to peck and tug and worry the young bird; but it is so stupid that it cannot understand.   Hours are spent in this way, and little progress is made ; at last the old bird dives down and comes up under the young one, which is nicely poised on its back.   In this way the mother swims off to sea with its offspring, rising and falling with the heaving of the billows until they are lost to view in the distance.   This is no mere hearsay—it is my own actual observation ; but my wife was the first on the island to observe it."[3]

In passing, it may be noted that, if the razorbill carries its

[1] Aquatic Birds, pp. 470-1.
[2] Vögel Mitteleuropas, xii. 167.
[3] Irish Naturalist, 1899, 135 (E. M'Carron)

offspring on its back out to sea, it might well, on occasions, adopt the same method in carrying it from the cliff to the sea.

That the young travel immense distances away from their natal cliffs is shown by the fact that a "guillemot marked as a newly hatched chick on the Aberdeenshire cliffs on July 11, 1910, was shot on November 29, 1910, a dozen miles north of Gothenburg, Sweden. This bird was thus four and a half months old when it was killed, more than 500 miles due east of its birthplace."[1]

The old continue to frequent the cliffs till August, when they leave for the winter, to lead a roaming life on the open sea. Their chief enemies at this time are heavy gales, which have the effect of driving away the fish, and so rendering the birds less capable of battling with wind and waves. Many are driven ashore, the beach being sometimes strewn with their bodies for miles. It is remarkable that, when amid the breakers, they make no attempt to escape by flight, but go on diving through the incoming waves until exhausted, when they are ruthlessly tossed ashore to feed the Gulls and Crows.[2]

THE GREAT-AUK

[F. C. R. JOURDAIN]

The opportunities of studying the life-history of this species have passed away never to return, for since 1844 no living specimen has been seen. The widespread but erroneous idea that it was an Arctic form, whereas in reality it inhabited the temperate North Atlantic, was responsible for vague surmises that in some unexplored part of the Arctic seas it might still survive. However, the interest which its disappearance from our fauna has added to the study of its

[1] Circular of the Aberdeen University Bird Migration Inquiry. "Bird-marking," March 1, 1911 (J. A. and A. L. Thomson).
[2] Saxby, *Birds of Shetland*, p. 285; Ussher, *Birds of Ireland*, p. 356; *Zoologist*, 1872, 2994, 3023; *Field*, 1863, 392; Yarrell, *British Birds*, iv. 55 (4th edit.); Dresser, *Birds of Europe*, viii., article "Razorbill."

history, and the careful sifting of the documentary evidence by such men as the late Professor Newton, Professor Wilhelm Blasius, and Symington Grieve, have resulted in a tolerably definite knowledge of its former range, and no hope can now be entertained of its existence in any part of the high north.

Meantime the literature of the subject has attained surprising dimensions. Wilhelm Blasius, in the new edition of Naumann's *Naturgeschichte der Vögel Mitteleuropas*, vol. xii. pp. 169-174, gives over five folio pages of closely printed titles of works bearing on the subject. The high prices reached in London auction-rooms by the sale of the eggs have no doubt helped to stimulate public curiosity on the subject, and perhaps the neglect and indifference through which this interesting bird was allowed to perish unnoticed, have been somewhat atoned for by the exceptionally careful way in which its history has been studied since its disappearance.

The most important papers in English on the subject are those by Professor Newton and Mr. Symington Grieve on its history, and Professor Owen on its osteology ; while Mr. F. A. Lucas and others have contributed to our knowledge of its American home. Among Professor Newton's contributions we may mention the articles in the *Ibis*, 1861, pp. 374-399; 1870, pp. 256-261 ; and 1898, pp. 587-592; his papers in the ninth edition of the *Encyclopædia Britannica* on "Birds," vol. iii. pp. 734-735, and "Garefowl," vol. x. pp. 78-80, which may also be found in the well-known *Dictionary of Birds*. The wonderful series of eggs and casts now in the Cambridge Museum are described in the *Ootheca Wolleyana*, vol. ii. pp. 364-383, and plates xiv.-xxi. Symington Grieve's most important work is his monograph on *The Great Auk or Garefowl*, published in 1885, to which supplementary notes have been subsequently issued by the same writer. Professor Owen's account of the skeleton of the great-auk will be found in the *Transactions of the Zoological Society of London*, vol. v. pt. iv. (1865) pp. 317-335, pls. li., lii., and map. Mr. F. A. Lucas has contributed papers to the *Auk* for 1888, the *Annual Report of the U. S. Nat. Mus.*, 1888-89, etc. But probably the most complete account is that of Professor Wilhelm Blasius, referred to above (pp. 169-208). Published in 1903, it contains practically all that was known on the subject up to that date. Since that time remains have been discovered in caves and kitchen-middens in various localities both in the British Isles and on the Continent, and it is only in this branch of research that we can hope for any further results.

As a British bird the great-auk was probably always a very local and far from common species. Putting aside the reported occurrences which are unsupported by actual proof,[1] the only evidence of its occurrence off the English coasts is furnished by the discovery in 1878 of a quite unmistakable upper mandible among extensive deposits of bones of various mammals and birds, as well as shells of molluscs, in a cave at Whitburn Lizards, at the eastern extremity of the Cleadon Hills, on the Durham coast. These remains were identified by the late John Hancock (see R. Howse, *Nat. Hist. Transactions of Northumberland and Durham*, 1880, vii. pt. 2, pp. 361-364). In Scotland remains are more numerous. Near Keiss, in Caithness, various bones were discovered in kitchen-middens in 1864, which were identified by Professor Owen (see S. Laing, F.S.A. Scot., *Prehistoric Remains of Caithness*, 1866). In 1879-81 Messrs. S. Grieve and W. Galloway explored a cone-shaped mound on Oronsay and obtained from it a number of bones of the great-auk, of which a complete list will be found in Grieve's monograph (see pp. 47-61). From this evidence Mr. Grieve infers that at one time the garefowl was common in the neighbourhood of Oronsay, and probably bred on the numerous rocky islets near its shores. It is probable that a careful and systematic exploration of the Hebrides would result in the discovery of many other remains.

There seems to be no doubt that St. Kilda was formerly a regular

---

[1] Such as those referred to in Hancock's *Catalogue of the Birds of Northumberland and Durham*, p. 165, and S. Grieve's monograph, pp. 24, 25. It is also said to have occurred near Lundy Island, on the coast of Cork, and in Belfast Lough (Yarrell, iv., edit. 4, p. 65). See also *Birds of Ireland*, p. 360.

breeding-place of this species. In 1680 Sir G. McKenzie refers to it as occurring very frequently and breeding there, and Martin, who spent three weeks on the island, says it arrives there regularly about the 1st of May, breeding on the bare rocks, and leaving about mid-June. Macaulay, who visited the island in June 1758, did not actually see the bird himself, but on the authority of the islanders states that "they do not receive annual visits from this strange bird, as from all the rest on the list," from which it is clear that in his time it was already much reduced in numbers. From this point onward its diminution must have been rapid, for the last record from this station is that furnished by Dr. Fleming (*Edinburgh Philos. Journal*, x., 1824, p. 94). The bird in question was captured during the early summer of 1821, by two young men and two boys, who were in a boat on the east side of the island and saw it sitting on a low ledge of the cliff. The two men then landed at opposite ends of the ledge, while the boys rowed the boat to beneath the rock on which the bird was sitting. Thus hemmed in on all sides, the bird jumped down towards the sea, but fell into the arms of one of the boys, who managed to hold it. The men then parted with the bird to Mr. Maclellan, the tacksman of Glass or Scalpa, who gave it to Dr. Fleming on the eve of his departure on 18th August from Scalpa, while on a tour of inspection in the yacht of the Commissioners of Northern Light-houses. "The bird," says Dr. Fleming, "was emaciated, and had the appearance of being sickly, but in the course of a few days became sprightly, having been plentifully supplied with fresh fish, and per-mitted occasionally to sport in the water, with a cord fastened to one of its legs to prevent escape. Even in this state of restraint it per-formed the motions of diving and swimming under water with a rapidity that set all pursuit from a boat at defiance." This bird is believed to have made its escape near the entrance to the Firth of Clyde while taking exercise in the sea.

From the Shetlands there seems to be no evidence of its occurrence, though it is said to have been seen on Fair Island. In the Orkneys, however, we meet with it again, but only in small numbers, though probably breeding. Low, who died in 1795, writes that he could find no evidence of its presence in the Orkneys, but when Bullock visited the islands in 1812, he was told that for several seasons a male and female had regularly visited Papa Westray. The female had been killed shortly before his arrival, and Bullock made desperate attempts to capture the male, chasing it for many hours together in a six-oared boat. The auk was so quick in diving, and moved so fast under water, that though they frequently got near him, they were quite unable either to shoot or capture him, and the pursuit had to be abandoned. However, subsequently some fishermen managed to catch the unfortunate bird, which was killed and the body sent to Mr. Bullock, by whom it was in due course transferred to the British Museum. Professor Newton's researches [1] tend to prove that the breeding-place of these birds was not on Papa Westray, but on the Holm, where there are sloping slabs of rock shelving gently up from the sea, sheltered from the winds and accessible even to flightless birds at all states of the tide.

The Irish records still remain to be considered. Until comparatively recently nothing was known of the occurrence of this species in Ireland beyond the immature bird captured in 1844. In the kitchen-middens of White Park Bay, on the Antrim coast, Mr. Knowles found in 1891 and 1895 numerous bones of great-auks, associated with flint implements and shells of edible molluscs. Some were scattered on the surface, and others were obtained by digging (see *Proc. Royal Irish Academy*, i. No. 5, p. 625 ; iii. No. 4, p. 654 ; and *Irish Naturalist*, 1899, p. 4). Similarly among the sandhills on Tramore Bay, Co. Waterford, Mr. R. J. Ussher found, in 1897, in different places among the extensive kitchen-middens, no fewer than seventeen bones of the great-auk, representing at least six individual birds. There is no doubt that the birds were used as food by the people who made the kitchen-middens, and it is scarcely possible that so many

[1] *Ibis*, 1898, p. 587.

could have been obtained unless some breeding-place existed in the vicinity. Probably, as Mr. Ussher suggests, the Waterford birds were obtained from the low-lying Keragh Isles, though the encroachment of the sea may have washed away any low islet then existing in Tramore Bay. Rathlin Island may well have afforded similar hunting-grounds to the dwellers on the Antrim coasts. Still more recently, remains have been recorded from the coasts of Clare, and among the sand-dunes at Rosapenna Mr. R. J. Ussher and Miss Weir found five humeri in July 1910.[1] Only one occurrence of this species within historic times is on record, namely, in May 1834, when a fisherman named Kirby managed to entice an immature bird to close quarters by throwing sprats to it, some miles to the west of Waterford Harbour. It was then captured in a landing-net, and appeared to be half starved. For some time after it would take no food, but potatoes and milk were then forced down its throat, after which it ate voraciously. It was fed chiefly on fish, which were swallowed entire, and trout were preferred to sea fish. "This auk stood very erect, was a very stately-looking bird, and had a habit of frequently shaking its head in a peculiar manner, more especially when any food was presented to it: thus if a small trout was held up before it the bird would at once commence shaking its head."[2] This specimen was fortunately preserved, and is at the present time in the Museum of Trinity College, Dublin.

Most of what little we know of the breeding habits of this bird will be found in the "Classified Notes," and need not be repeated here. Although in some cases pairs appear to have bred on isolated stations, as in the Orkneys, we know that when the bird was common, as on Funk Island off the Newfoundland coast, very large numbers were found breeding close to one another. No nest whatever was made, the single egg being laid on the dung-covered rock. From the fact that "brooding spots" have been found in skins of both sexes, it

[1] See *Irish Naturalist*, 1910, pp. 171 and 195, pls. 10, 14.
[2] Dr. Burkitt, quoted in the *Birds of Ireland*, p. 359.

has been inferred that both male and female took part in incubation. Some of the eggs now in existence show traces of a bluish green ground-colour. Naturally in the course of years this colour would have faded considerably, especially as many of them were kept as ornaments and constantly exposed to the light. In this connection it is interesting to note that Martin speaks of the eggs as "variously spotted, black, green and dark." Probably when fresh they resembled gigantic eggs of the razorbill, some showing a tendency to approach the ordinary type of the guillemot in colour. At the present time it is believed that there are about 73 eggs in existence, and probably about 77 skins or mounted specimens.

The only sound which we know this bird to have made is a low croak. They seem to have been inoffensive, never making the slightest attempt to defend their eggs, but occasionally biting fiercely when handled incautiously. From the fact that two specimens are known to have been captured after being enticed to within a yard or two of a boat, it is clear that these birds had no great fear of man. In fact, all that we know of them seems to show that their intelligence was of a low order, and probably only their extraordinary powers of diving and swimming saved them from disappearing at an earlier period of history from the Atlantic.

BLACK-GUILLEMOT

[F. B. KIRKMAN]

The black-guillemot is not, as its English popular name implies, a member of the same genus as the common guillemot; it differs from the latter both in its internal and external structure. The most obvious external differences are in the shape of the bill and the coloration of the plumage. In the breeding season the guillemot has a somewhat Pickwickian aspect, owing to its expanse of white breast. The plumage of the black-guillemot, on the other hand, is, at the same period, greenish black all over, except on the wing, where there is a bright white patch, which is conspicuous even when the bird is under water. But the chief glory of the species is the vermilion-red of the legs and of the inside of the mouth, which are in marked contrast to the dark plumage.

The black-guillemot, unlike the other British breeding Auks, is not found in large colonies, but this appears to be due rather to its scarcity than to its objection to the society of its own kind. Naumann's statement that it is gregarious, but not usually found together in larger numbers than from twenty-five to thirty pairs, is correct as far as the evidence goes.[1] It is frequently found in smaller numbers. In Ireland, for instance, as a rule, "one or more pairs only are met with in the same place."[2] The relative scarcity of the species is difficult to account for, the more so as, unlike its British breeding congeners, it lays 2-3 eggs instead of only one. On the other hand, it appears seldom to rear more than one chick.

The black-guillemot is less inclined to consort with other species than are puffins, guillemots, and razorbills, but all four species may be occasionally found nesting together.[3] It is seen on the waters near its breeding-haunts towards the end of February or in March, but it does not appear to be in regular occupation of its nesting-place till April.[4] Naumann states that the birds arrive in pairs.[5] Like their congeners, the mated birds show their affection for each other by billing and other caresses, which have yet to be exactly described, and, according to one observer, by "dignified bowings."[6]

The regular pairing call of the species is a clear "*ist, ist, ist*," according to Naumann, who also ascribes to it a whistling note.[7] The latter is probably the same as the "low plaintive whine,"[8] or the "shrill, rather faint whistle"[9] of other writers. According to Mr. Selous, "the

[1] *Vögel Mitteleuropas*, xii. 239.   [2] Ussher and Warren, *Birds of Ireland*, p. 366.
[3] Patten, *Aquatic Birds*, p. 491 ; Naumann, *Vögel Mitteleuropas*, xii. 239.
[4] Cf. *Ann. Scot. Nat. Hist.*, 1904, 23 ; Ussher and Warren, *Birds of Ireland*, p. 366 ; Naumann, *Vögel Mitteleuropas*, xii. 240.   [5] *Loc. cit.*
[6] Job, *Wild Ways*, pp. 174-80.   [7] *Op. cit.*, p. 238.
[8] Dresser, *Birds of Europe*, viii.   [9] *Loc. cit.*

cry is, for the most part, a weak, twittering sound, but occasionally rises into a very feeble little wail or scream. All the while the bird is uttering it, he keeps raising and again depressing his head and opening his beak so as to show conspicuously the inside of his mouth, which is of a very pretty rose or blush-red hue, almost as vivid as that of the feet." [1]

Whether the cock feeds the hen is not recorded. Both share in the incubation of the eggs, [2] which are laid on the ground, no nest being built. Both, again, share in feeding the young. "They bring, each time, a single fish—a sand-eel, often of a fair size—and disappear with it into the hole, reappearing shortly afterwards. . . . What I particularly noticed was that when the bird that had taken a fish in had come out again, the other, even though it had nothing, would always go in too, as though to pay the chick a little visit. It stayed about the same time—less than a minute that is to say." [3]

From this account, and Naumann's, [4] it appears that the young are fed on fish. The diet of the parents, as already noted, is said to consist chiefly of crabs. In their method of pursuing their prey the species does not seem to differ from the guillemot and other Auks, unless it be that they make more use of the legs, which appears to have been the opinion of Saxby. He states that after diving beneath the surface, they move "quickly from rock to rock by the help of feet and wings, turning aside the long weeds with their bills, and after each capture of fish or crustacean, coming to the surface to complete the process, seldom making another dive without half-rising from the water, and flapping the wings to free them from superfluous moisture." [5] According to Mr. E. Selous, the wings only are used. Saxby may have meant that the feet are used in turning, and this may be true of the other Auks.

Saxby goes on to mention that "for some time after the descent

---

[1] *Bird Watcher in the Shetlands*, p. 128.
[2] Saxby, *Birds of Shetland*, p. 297; Naumann, *op. cit.*, p. 240. See also "Classified Notes."
[3] E. Selous, *Bird Watcher in the Shetlands*, p. 72.
[4] *Op. cit.*, p. 240.   [5] *Birds of Shetland*, p. 297.

Plate 96

Upper: Black-guillemots and young
Lower: Little-auks in winter plumage

By A. W. Seaby

is made, the numerous air-bubbles clinging to the plumage give the bird a singular but remarkably beautiful appearance.[1] The average duration of each submergence is as nearly as possible twenty-four seconds, the bird making a slight sudden splash with its wings as it disappears." He adds that it dives so quickly that it is not easy to shoot on the water. Instead of diving, it often runs along the surface, flapping its wings, and striking the water with its feet more quickly than a coot.[2]

The black-guillemot differs from the other British breeding Auks not only in laying two eggs, but in that the young quit the nest only when able to fly,[3] that is, leave it for the water, for, when a fortnight old, they will come out of the holes and stand at the entrance to receive food.[4] How long the young continue to be fed after they have taken to the water is uncertain. So also are the winter movements of the species. They are said to be less oceanic in their habitat than their congeners; they prefer sheltered waters along the coast to the open sea.[5] According to Naumann, they suffer considerably during stormy weather, many being washed ashore dead. But in British waters their bodies are rarely found among those of their congeners tossed up on the beach, a fact which emerges very clearly from the lists of casualties published after the great storms of February 1872.[6]

[1] The same applies more or less to most, if not all, diving birds.
[2] Ussher and Warren, *Birds of Ireland*, p. 366.
[3] Patten, *Aquatic Birds*, p. 490.
[4] Naumann, *Vögel Mitteleuropas*, xii. 240.
[5] Patten, *Aquatic Birds*, p. 490; but cf. Naumann, *op. cit.*, p. 240, who states that, while the young of the year remain in the sheltered bays all the winter, the old remain on the open sea, except when driven in by storms.
[6] *Zoologist*, 1872, 2904, 3023. See also Patten, *Aquatic Birds*, p. 490.

THE LITTLE-AUK

[F. C. R. JOURDAIN]

Although practically resident in the Arctic Ocean, from Baffin's Bay eastward to Novaya Zemlya and the North Atlantic, moving somewhat northward during the breeding season and southward in the winter months, this species occurs in most years in varying numbers on our coasts. Sometimes only a few individuals are recorded from the northern outposts, such as the Shetlands and Orkneys,[1] while again at other times a great body of southward-bound migrants pours into the North Sea and along our east coasts. Very few comparatively find their way along the western side of Scotland, but occasionally they have been recorded in fair numbers, though not approaching those met with on the east. One of the earliest invasions of which we have definite information is that of 1841, when, after several days of stormy weather, great numbers appeared off the Northumberland and Durham coasts, as well as off Redcar in Yorkshire, while smaller numbers found their way along all the eastern counties south to Kent and Sussex. There is good reason to believe that in favourable winters these auks can subsist in the northern seas without finding it necessary to resort to land at all. A succession of stormy days, however, prevents them from feeding, and in their weakened condition they are driven exhausted and starving on to the nearest land. Many in this state are drowned, but some take to flight and drift long distances inland.

Another visitation to the north of England took place in 1863, when large flocks appeared off the Durham coast and at the mouth of the Tees, and since that date several other important irruptions have taken place. Of these, perhaps the most extensive was that of 1894-95. In January 1895 the weather was extremely severe, accompanied by snowstorms, and thousands of these unfortunate birds were observed on their way south, drifting before the storm. Many were

[1] Cf. note on *Migration*, p. 10.

completely exhausted, and were picked up either dead or in a dying condition, after their long struggle against wind and waves. Near Redcar alone two hundred and fifty are known to have been found,[1] and two hundred and eighty-five from Norfolk,[2] but this can only represent a small proportion of those which perished. One peculiarity about this invasion is obvious from a consideration of the Scottish records, namely, that though very few birds occurred on the west side of Scotland north of Oban, a considerable number evidently found their way down the Great Glen, no fewer than twenty-six having been picked up in the environs of Oban itself. The irruption of February and March 1900 was also on a large scale, but was chiefly noticed along the Norfolk coast from the Wash to Lowestoft. Here the numbers probably exceeded those of 1895, but the invasion was confined to a smaller area. Of these three or four more important irruptions, it is probable that that of 1841 was the most extensive, while that of 1895 was also on a very large scale. The Scottish records of 1895 will be found carefully studied and digested by Mr. W. Eagle Clarke in the *Annals of Scottish Natural History*, 1895, pp. 97-108, and by Mr. J. Paterson in the *Proceedings of the Natural History Society of Glasgow* for the same year.

Apart from these occurrences, which are directly due to stress of weather during the winter months, there is no reliable information as to its presence with us at any other season,[3] though a few examples have been obtained in summer plumage. Its breeding haunts lie far to the northward, and it has not been known to nest even on the Færoes, while on Iceland it is practically confined to Grimsey, though possibly a few pairs may breed on the mainland.

The Grimsey colony is not a large one, apparently consisting only of about one hundred and fifty to two hundred pairs, but farther north it becomes more and more numerous. Enormous colonies of many hundred thousands of pairs exist in Spitzbergen, Franz Josef Land,

---

[1] *Zoologist*, 1895, p. 68.   [2] *Ibis*, 1896, p. 276.
[3] One in full plumage was, however, obtained in the Monach Isles, Outer Hebrides, on June 24, 1893.

the Greenland coasts, etc. The time of arrival at the breeding-places seems to vary somewhat, and is recorded at different stations from February 25 (Nansen), March 2 (Dr. Neale), and March 9 (W. S. Bruce), while Pike did not notice it on Spitzbergen till March 28. Rotches are extremely sociable birds, and are almost always to be met with in flocks. When they reach their breeding-places they seem to observe regular hours, setting out for the open sea to feed in a steady stream, and returning later in broken lines to the rocks. During April, May, and early June, W. S. Bruce notes that the breeding cliffs were sometimes thickly tenanted, and again at other times apparently deserted. After June 10, however, serious breeding operations began, and from this time onward the rotches were regularly seen on the cliffs. The breeding birds are accustomed to travel considerable distances in search of food, so that one comes across flocks busily engaged in feeding many miles away from the breeding-ground. Chapman describes them as swimming rather deep in the water, and " by the stern," while the Rev. A. E. Eaton gives an amusing description of the return of a feeding-party. "The rotche when it flies has always the appearance of being rather behind its time : it seems in such a tremendous hurry, and starts off with its mouth crammed full of food, as if it had been suddenly called away in the middle of dinner. You may see a party of them on the water—six or seven birds—take wing together to return to their nests. You think they are all gone, but you are wrong : for without pausing for an instant to see whereabouts they are, Nos. 8, 9, and 10 come flying up from under water, one after the other, and take after the others at full speed."[1] As will be seen from the above extract, the rotche is an expert diver : in fact, he is somewhat disinclined to trust to his wings when disturbed by the approach of a ship, and has a knack of splashing and scuttering along the surface of the water, as if unable to fly, till he meets an approaching wave, and promptly dives into it.[2] Enormous flocks are sometimes to be met with : a single shot has been known to bring

[1] *Zoologist*, 1874, p. 3819.　　　　[2] Saunders, *Manual*, p. 706.

down no fewer than thirty-two birds; and a party of three guns sent out to procure food killed no fewer than twelve hundred and sixty-three birds in five or six hours in mid-August.

Under water rotches use their wings just as in flying, and Feilden notes "that the individuals in a diving flock kept their relative distances and bearings under water with as much correctness as if on the wing, and all returned to the surface within a second of one another. During the breeding season the pouch-like enlargement of the cheeks gives them a singular appearance. The contents of the cheeks is a reddish-coloured substance, which on closer examination is found to consist of immense numbers of minute crustacea." Colonel Feilden goes on to point out that the guillemots, razorbills, and puffins, which live on fish, have no difficulty in transporting food to their nests, but in the case of the rotche the bill is useless for this purpose, while the breeding-grounds are in some cases far inland, so that some structural modification is necessary to enable it to transport food in sufficient quantity, and this has been obtained by the distention of the cheeks.

At the breeding-grounds the food of these birds consists practically entirely of small crustacea (*Entomostraca*). The red colour of these minute organisms stains the droppings of the bird and causes them to assume a vermilion tint, conspicuous against the snow, and sometimes giving the first indication of their arrival.

The little-auk is a noisy bird, restless in its habits, and seldom still. On the cliffs Hantzsch describes it as continually moving a step or two and uttering its peculiar call. Malgrem says that the neighbourhood of a breeding-place can be detected half a mile away by these notes. The call has been variously written, "*Rett, tet, tet, tet*," or "*Perre, te-te-te, tett, tett, tett*," and "*Trrr, trrr, tet, tet, tet, trrr*." Le Roi describes the noise as somewhat resembling "*Prrrrrrr quiequiequie-quiequie*," and as it is uttered by hundreds of birds almost continuously, it is naturally audible for some considerable distance. For its nesting-place it chooses very similar sites to Mandt's guillemot. Sometimes the two species may be found breeding together, though in separate colonies. As a rule, the eggs are difficult to get at: they are laid on the bare rock; but deep down in crevices of the cliffs, or underneath masses of boulders which form the talus. Many of the cliffs where the rotche breeds are quite inaccessible, and even where the holes are low down the friable nature of the rocks often makes climbing dangerous, or the egg is so far down that it cannot be reached by hand. By these means it manages to defy the Arctic foxes, which are continually prowling about the foot of the cliffs. Many auk cliffs are some distance from the sea. On Spitzbergen, while Messrs. Trevor-Battye and Garwood were descending the well-known Horn mountain on August 17, they met with a single little auk at a height of about 3000 feet. It flew round and round, but just below them as they looked down over an *arête*, evidently anxious about its young, and uttering a twittering note. Only a single egg is laid as a rule, but Mr. C. Ingram found two fresh eggs in one hole in Spitz-bergen, while another was tenanted by two young about the same age. It is of course possible that in both cases the single hole was occupied by two hens, but as most birds had young at the time of his visit, it is very remarkable that the two eggs were both fresh. Still, in any case, these must be regarded as rare exceptions, and the vast majority of birds lay one egg only. In colour it is a pale greenish blue, sometimes unmarked and at other times spotted and streaked with light rust colour, chiefly at the big end. The greenish ground-colour tends to fade, leaving the shell a greenish white. The average size of 92 eggs is 1·92 × 1·33 in. [48·7 × 33·9 mm.]. In Iceland the first eggs may be found at the end of May, but the regular breeding time is in early June, while in Spitzbergen probably the best time is about June 18-21. Mr. C. Ingram states that though the nests contained young, closely huddled rows of auks were sitting on ledges of the cliffs, while others were flying round in small parties of ten to twenty or more. While on the wing they were continuously giving vent to weird laughing cries, but when close at hand they conversed together with low, marmot-like noises. "When close to their nests they would come and circle very near to my head, checking their flight by holding out their webbed feet, and making short wing strokes."[1]

On account of the situation of the nest, and the inaccessibility of its breeding haunts, it is naturally difficult to obtain accurate data as to the length of the incubation period, but Hantzsch estimates it at about twenty-four days. There is no doubt that both sexes share in the duty of incubation, for not only are breeding spots found in both males and females, but Dr. Le Roi caught many of both sexes on the eggs. From early July onward the young may be found in the nesting-holes, tiny little creatures, covered with uniform sooty or almost black down. One taken by Mr. Ingram ate raw fish and meat with avidity, after food had once been forced down its throat, and learned to call eagerly whenever it heard him speak. He remarks that it had a very curious habit of wagging its head from side to side after every mouthful.[2] At a later period, when half fledged, the under parts are white, according to Saunders.

By the end of August most of the breeding-places have been deserted by the young, which have now taken to the open water. The larger gulls and the foxes have, however, taken toll of them on their way. They do not disappear altogether from their most northerly breeding-places till about the first or second week in September, while Arnold Pike notes their last appearance in his winter diary on Spitzbergen under the date October 13. Gradually the great flocks move southward before the approach of winter, but as with the millions of other Arctic rock birds, such as Brünnich's guillemot, Mandt's guillemot, Arctic puffin, etc., we have no definite information as to how or where they spend the winter months. Somewhere or other in the sunless North Atlantic these hosts of birds must winter, but even the thousands of storm-driven rotches which occasionally

[1] *Avicultural Magazine*, New Series, vol. iii. (1894-95) p. 358.
[2] Cf. Dr. Burkitt's account of the behaviour of the great-auk captured alive off the Waterford coast, quoted in the *Birds of Ireland*, p. 359.

drift down our eastern coasts can form but an infinitesimal fraction of the vast multitudes which must spend the Arctic nights in as yet unknown reaches of the northern seas.

THE PUFFIN

[F. B. KIRKMAN]

The puffin is one of the most singular birds that make their home on the ocean. Its singularity, if one considers it well, seems to lie in the striking contrast between the gravity of its demeanour and the fantastic shape and harlequin hue of its beak. As it sits on the top of some rock, presenting its white breast to the sea, it has somewhat the appearance of a pantomime Napoleon with a very large and richly coloured false nose. The puffin's gravity and its beak or neb are indeed responsible for most of the popular names which have been bestowed upon it, such as pope, bottle-nose, and sea-parrot. The shape of the beak has given it another name, coulterneb, for it is supposed to resemble a coulter, the fore or cutting iron of the plough.

In the air the bird has a more graceful appearance than either of the Auk species—guillemots and razorbills—with which it consorts. This is due to its markedly longer wings.[1] It sometimes whirls about with a lightness and agility that is almost deceptive. Like its congeners it progresses with rapid strokes—its narrow wings vibrating rather than flapping. Not that the Auks always fly with rapid beats. I have seen both razorbills and guillemots quit the ledges with slow measured flaps very unlike their ordinary movements, but am not certain whether the puffin does the same. The periods in its flight

---

[1] Razorbill—length 17 inches, wing length 7·3; guillemot—length about 18 inches, wing 7·5; puffin—length 13 inches, wing 6.

when it is neither graceful nor imposing are those when it is quitting the land or about to return to it. Then the webbed feet hang down straddle-wise, and, as they happen to be large and unshapely, and of a bright orange hue that renders them conspicuous, the general effect is inartistic, not to say grotesque.

Puffins arrive on the waters near their breeding-haunts in March, but do not busy themselves with their nest-holes till April. In the meanwhile they appear to visit the rocks irregularly, or at least to be present on the waters close to them, and disappear at night to sleep on the open sea. But little is known of their habits at this period, apart from the mere dates of their arrival.

Nor is much known of their love-displays. As might be expected, the neb plays an important part in the proceedings; pairs may frequently be seen billing, also shaking their heads, nibbling each other's plumage, and making grave little bows. Occasionally the head is bent right down to the feet. The bird then seems to be peeping through his legs at the prospect behind, and presents a very comical appearance. When the guillemot bows to the ground, he appears, as already described, to be about to push an invisible egg under his breast. The difference is due to the fact that the latter species generally stands on the whole foot,[1] whereas the puffin, like the generality of birds, stands on his toes. This, it is true, has been disputed. The bird is said by Professor Patten to stand on the toes only when alarmed, "the position generally depicted in 'photographs from nature.'"[2] I should have put it the other way about: when alarmed and preparing to take flight the birds will naturally flex the leg in order to spring off, and so have the appearance of resting on the tarsus; otherwise they stand upon the toes, as I have frequently had occasion to observe from a distance, and through a strong field-glass, when they were not in the least alarmed, as evidenced by the fact that I noted individuals among them indulging

[1] See p. 27.
[2] Patten, *Aquatic Birds*, p. 499. See also for same error H. Saunders' *Manual of British Birds*, 2nd ed., p. 708.

in their love-displays.[1]   The species also walks on the toes with a springing active movement.

The puffin is a comparatively silent species, and is often content merely to open and close its particoloured beak, leaving the sound to be taken for granted—a praiseworthy economy of energy, having regard to the unmusical character of its utterances.   These have been described as a long-drawn, very grating "*owk*" or "*ow*," and a long-drawn "*oooo*," something like the nocturnal call of a cat.[2]   What meanings are to be attached to each have yet to be precisely determined. The young bird, when handled, utters a piping note syllabled as "*yerp*."[3]

Of all places for its nest-hole the puffin prefers the earth-clad tops or sides of isles or cliffs.   At the Farnes, in the Longstone group, for instance, it occupies the upper surface of one of the islands, leaving the bare tops of the adjacent rocky stacks, known as the Pinnacles, and the ledges on the steep sides of the island, to the razorbills, guillemots, and kittiwakes.   Similarly, the huge colony at Mingulay in the Outer Hebrides nests in the earth on top of the stack of Lianamull, beneath a dense crop of seeding sorrel, which later on becomes, however, one sticky compound of mud, dung, feathers, rotten eggs, and dead birds, ankle deep or deeper.[4]   Another nesting-place, on one of the low-lying Scilly Isles, is shown on Plate XLI.   The nest is also to be found in the holes or crevices in cliffs, or under loose stones or boulders.   A photograph of the latter, taken at one of the North Farnes, is shown on the Plate above mentioned.

The holes in the earth may either be made by the puffin itself or be that of a rabbit.   If the latter is in occupation of a hole that the puffin has set its heart upon, there is a forcible eviction.   The young

[1] This view is corroborated by Mr. C. J. King, who in the course of twenty years' watching of puffins on the Scilly Isles never saw them rest on the tarsus except when preparing to take flight (*in litt.*), by Mr. O. V. Aplin and others, and is, I believe, generally accepted.

[2] *Zoologist*, 1910, 41 (O. V. Aplin).

[3] O. V. Aplin, *loc. cit.*

[4] Harvie-Brown and Buckley, *The Fauna of the Outer Hebrides*, p. 166.

rabbits, if there are any, are removed one by one by the ear or neck, and deposited outside. Then follows a struggle with the old rabbit, who has usually to yield to its adversary's beak. It is "bitten" out.[1]

In excavating its own hole, the puffin uses its feet, scraping the earth out backwards, so that it looks like "a fountain of dust blowing up from the burrow."[2]

The labour of excavating a new hole is, of course, unnecessary in the case of most birds, as they return to their holes of the previous year, which they clean out and refurnish. The frequent disputes and fierce fights for holes which take place among the puffins themselves seem to show either that some birds are too lazy to excavate, or that there is not room for all. The fact that there are usually a large number of non-breeding birds points to the latter conclusion. At Lovunden in Norway, where there is a colony of some millions, the non-breeding are said by Collett to exceed the breeding birds. At another smaller colony the proportion of breeding to non-breeding is given as 1 to 20.[3]

Unlike the other British Auks, the puffin does not merely lay its eggs upon the ground, but usually constructs a nest of grass, roots, and feathers. According to Naumann, both sexes share in this labour. The one whitish coloured egg is usually laid in May. The bird can be drawn from its hole without resistance, and does not, in fact, appear to bite or scratch till it is out, and can see clearly what it has to deal with. Then it makes up for lost time.[4]

In June or later the chick, which is clad in thick jet-black down, makes its appearance and is assiduously fed by both parents on fish. These are caught by diving and swimming or flying under the water. In its sub-aquatic movements the puffin resembles the other

---

[1] Payne Gallwey, *The Fowler in Ireland*, p. 265. A detailed account of an eviction has yet to be written.
[2] F. Heatherley (*in litt.*).
[3] Naumann, *Vögel Mitteleuropas*, xii. 254.
[4] Cf. Naumann, *Vögel Mitteleuropas*, xii. 255-6 ; *Zoologist*, 1901, 146.

auks ; it uses its wings, the legs being trailed behind, as in flight. Whether the legs are used to help in turning or steering is not clear from the evidence. I have personally seen them trailing only. In its method of carrying the fish the puffin resembles the razorbill ; it brings to the nest half a dozen or so at a time, the heads and tails forming a fringe on either side. The guillemot and the black-guillemot carry one fish held lengthwise. To catch and hold one fish is easy enough, but how are the puffin and razorbill able to catch the second fish without losing the first, or the third without losing the other two, and so on ? [1]

One explanation may be found in the fact that the palates of the razorbill and puffin are furnished with rows of little pointed spines directed towards the throat. Mr. C. J. King of St. Marys, Scilly, first kindly drew my attention to the presence of these spines on the palate of the puffin. It occurred to me that they ought to be present on the palate of the razorbill. This I found to be the case, but I found it also to be the case with the guillemot, which has no need of such an apparatus, as, unlike its two congeners, it only catches and carries one fish at a time. Further, I noted that all three species had the spines at the posterior end of the tongue, and Mr. Pycraft, to whom I sent the specimens, found that they had much larger spines a little way down the throat, on either side and at the back. Neither here nor at the back end of the tongue could the spines be of use for holding fish carried in and across the bill. Palatal spines are found in nearly all birds, though in some they are very small. It is the absence, rather than the presence, of these excrescences that should excite comment. Thus they seem to be absent among Accipitres, but present in the Striges—birds with precisely similar feeding habits. In the penguins, some species of which feed largely on minute crustacea, others on fish, they are of huge size, especially on the tongue.[2] That the spines have been developed for the purpose of giving a firm

---

[1] The usual length of the fish caught by puffins is three inches, but they may vary in size from 2 to 6 inches.—*Zoologist*, 1910, 41 (O. V. Aplin).   [2] W. P. Pycraft (*in litt.*).

grasp upon fish is, therefore, very improbable. That they have been developed for some other reason, and yet assist, in the case of the puffin and razorbill, to maintain the fish in place, is an arguable position ; but even then it has yet to be proved that the spines are large and strong enough to prevent a fish becoming detached from the palate and upper mandible when released from the grip of the bill. A more plausible theory is that the fish are held in place by the tongue each time that the bill is opened. This is probably also what happens in the case of birds, like Thrushes or Wagtails, which are able to seize and hold yet another insect or worm without disturbing the bundle already in the bill. As these species are without palatal spines, it would be difficult to suggest any other explanation. And if the explanation suffices for them, why should it not suffice for the puffin and razorbill ? The palatal spines must, however, serve some purpose. What this is may well remain an open question.

Whether the external shape and the size of the beak is of aid in giving a firmer grip upon the fish is not yet clear, but in this connection it is instructive to note that the beak of the puffin and razorbill are much alike in shape, and differ considerably from that of the guillemot ; also that during the breeding season, that is at the time when most fish have to be caught, the beaks of the first-named species are actually larger in size than in the winter. This remarkable fact was discovered, in respect to the puffin, by a French zoologist, Dr. Bureau.[1] He showed that after the breeding season, in August, the basal half of the beak, the pieces constituting the blue area, and the yellow line bordering the outer edge of the blue (Fig. 1, B), scale off in plates. I have, in fact, been able to lift easily with a penknife the blue plate on the upper mandible of a freshly killed specimen examined on June 22. Underneath it was a fleshy layer or pad, which presumably hardens to form

[1] *Bulletin de la Société Zoologique de France*, vol. ii. 377-399. Translated in part in the *Zoologist*, 1878, 233. The same observer found that similar changes take place in other species of puffin and of related genera, for which see the *Bulletin Soc. Zool. France*, vii. 270 ; viii. 348.

the surface in winter. With these plates disappears the horny, whitish, curiously indented fillet along the base of the upper mandible (Fig. 1, *F*), while at the same time the yellow rosette (Fig. 1, *Y*) at

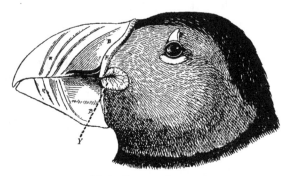

Fig. 1.—PUFFIN IN SUMMER.

the gape becomes shrivelled and discoloured, and the appendages above and below the eye disappear. The front half of the beak, consisting of the red triangular end, the red ridges and three furrows,

Fig. 2.—PUFFIN IN WINTER.

remains unaltered (Figs. 1, 2, *R*). On examining skins of razorbills at the Natural History Museum, South Kensington, I found that this species also showed in winter a distinct diminution in the size of the posterior half of the bill, the part, that is, behind the median white transverse line (Figs. 3, 4). On the lower mandible, the part of the beak next to the chin is visibly smaller, the result being to make the underside of the mandible come more distinctly to a point in its middle. The fillet along the base of the upper mandible disappears, as in the case of the puffin (Fig. 3, *).

Fig. 3.—RAZORBILL IN SUMMER.

Having caught its fish, the puffin carries them to its hole, and is said to drop them before the young, and gives them to it one by one. Macgillivray states that the newborn chick is fed by regurgitation, or receives little pieces of fish which are placed in its mouth, and that the larger pick up the pieces from the ground at the entrance of the hole, but he does not make clear whether he saw this take place or accepted it on hearsay.[1]

Fig. 4.—RAZORBILL IN WINTER.

When, in July or August, the young are ready to quit their hole they go fluttering, rolling, running, and tumbling down to the sea.[2] This is said to occur chiefly at night or daybreak.[3] Of their behaviour on reaching the water, and the time of their departure, too little is recorded to be worth noting. At the end of August, or later, both young and old have disappeared out to sea. They are found in winter over the North Atlantic and in the western Mediterranean.

The chief enemies of the puffins, apart from the gales which toss them dead ashore, are the hawks and larger gulls. The peregrine has been seen to snap up the young when they are old enough to appear at the entrance to their holes for food.[4] The remains of old and young may be found all about the colony. The former are often seized unawares as they are issuing from their nest-holes. Mr. C. J.

[1] *History of Birds*, v. 372.
[2] Ussher and Warren, *Birds of Ireland*, p. 369 (M'Carron) ; Patten, *Aquatic Birds*, p. 500.
[3] Ussher and Warren, *loc. cit.*   [4] Ussher and Warren, *Birds of Ireland*, p. 369.

King writes me that the lesser blackbacked-gulls " watch at the mouth of the nesting-hole, and, as the puffin comes out, lay hold of it by the back of the neck, shake it like a terrier shakes a rat, and when they have killed it, disembowel it and leave the empty carcase. A year or two ago," he continues, " I collected within a radius of about fifty or sixty feet a heap of about thirty of these victims." Herring-gulls and lesser blackbacks have been seen to push their heads down the holes and pull out the young, and the great blackbacked-gull will boldly seize an adult puffin from among those in the midst of which it happens to be standing. Of this more will be said in the chapter on these Gulls.

THE TERNS

[ORDER: *Charadriiformes*. FAMILY: *Laridæ*. SUBFAMILY: *Sterninæ*]

PRELIMINARY CLASSIFIED NOTES

[F. C. R. JOURDAIN.   F. B. KIRKMAN.   W. P. PYCRAFT.   A. L. THOMSON]

BLACK-TERN [*Hydrochelidon nigra* (Linnæus). Starn or stern, black-kip, coal-kip, blue-darr, carr-swallow, carr-crow. French, *guifette noire*; German, *schwarze Seeschwalbe*; Italian, *mignattino*].

1. Description.—The black-tern is to be distinguished from its congeners, when in its summer dress, by its pale grey under wing-coverts, upper tail-coverts and tail, and black bill; in winter plumage it resembles the whiskered-tern, having the rump grey, like the back, but it differs from the whiskered-tern in its smaller size, more slender bill, and in having the webs of the toes less deeply encased. (Pl. 98.) The sexes are alike. Length 9·5 in. [241·30 mm.]. In summer the head, neck, and breast are black, the rest of the upper parts dark slate-grey, paler on the wing-coverts, which along the carpal bend of the wing are white; the shafts of the primaries are white. The upper tail-coverts and tail feathers are of a pale slate colour. The under parts are leaden grey, the hinder flank and under tail-coverts white. Beak black, feet and toes reddish brown. After the autumn moult the forehead is white, the hinder crown and nape black but "frosted" with white, and the side of the face, neck, a collar round the hind-neck, and the under parts are white, more or less intermixed with grey. In the juvenile (fledgling) dress the crown, sides of the face, and interscapulars are sooty black, in some cases slaty black; the wings light slate, the coverts being tipped with dark brown, while the whole of the exposed portions of the inner secondaries are of a dark brown; the primaries have each very narrow white lines along the inner web, and the tail feathers are tipped with brown. The under parts are white, but the upper flanks are slate-grey. This is succeeded by a plumage in which the mantle and wings are French grey, strongly washed with brown and with broad coffee-brown margins to the interscapulars and scapulars. Marginal wing-coverts black—white along the free edge of the anterior border—the rest of the coverts and quills dark grey. The forehead is white, washed with rust colour, crown and sides of the head slaty black with a tinge of brown ; the hind-neck is grey, the sides of the neck white, washed with rust colour, the rest of the under parts being white, save the anterior flank-feathers which are slate-grey. The young in down is of a rich buff above, with broad conspicuous black markings representing disintegrated median and lateral longitudinal stripes. The pattern on the head forms a more or less complete black ring on the crown, a longitudinal stripe on the nape, bounded on either side by two black spots. The fore-neck is white washed with brown, and the breast and abdomen are dull white. [w. p. p.]

2. **Distribution.**—Though formerly a common summer resident in the Fens and the Norfolk Broads, and reported as having bred in the Solway marshes, the black-tern has not been known to breed in England since 1858, when a pair nested at Sutton in Norfolk. At the present time the breeding range of this species on the Continent extends to Southern Sweden (north to Upland), the islands of Öland and Gotland in the Baltic, and Russia up to lat. 61° on the west side and 58° on the east. From these limits it is fairly general in marshy districts southward over the rest of the Continent, though very local in France, breeding in some numbers in Southern Spain, but only found in the marshes of Northern Italy, and absent from the south of the Balkan Peninsula. · It does not breed in the Mediterranean islands, except in the Balearic Isles, and there is no reliable evidence of its nesting in North Africa at the present time, while in Asia it visits Western Siberia and is found east to about 85° long., but not farther south than the shores of the Caspian. Its winter quarters lie in tropical Africa—on the west side Gambia, Liberia, Fanteeland, Kamerun, the Congo district, and Angola, on the east side it migrates through Egypt and Nubia to Kordofan, Abyssinia, British and German East Africa. In North America it is replaced by a closely allied form, *H. nigra surinamensis*, whose winter quarters extend southward to Chile and Peru. [F. C. R. J.]

3. **Migration.**—A bird of passage, presumably on its way from it winter quarters to the Baltic countries ; formerly also a summer visitor and a British breeding bird. In April and May it occurs in small numbers on the south coast of England and on the east coast south of the Humber; records from other parts are exceptional. It is also a casual visitor during the summer months, presumably from Holland. In August and September a more noticeable passage occurs,

PLATE XLII

Photo by W Farren

Black.tern's nest

Photo by F. B. Kirkman

Sandwich.tern's nests in marram grass at Ravenglass

consisting mainly of immature birds in small parties. At that season the birds are more widespread, occurring irregularly in Ireland, and they also penetrate some distance inland along the rivers. A few may linger as late as November in the south-west (cf. Saunders, *Ill. Man. B. B.*, 2nd ed., 1899, p. 633 ; and Ticehurst, *B. of Kent*, 1909, p. 490). [A. L. T.]

4. **Nest and Eggs.**—Although no longer one of our breeding species, the black-tern continues to visit us occasionally in the summer months. Its nest consists merely of a few water-weeds and decaying vegetable matter, built up till just above the level of the shallow water, and partly supported by rushes and other growing water-plants. (Pl. XLII.) The materials are provided by both sexes. The eggs are usually 3 in number, occasionally only 2, and are olive-green or ochreous in ground-colour, spotted and blotched with irregular markings of very dark sepia, almost black, and purplish grey underlying shellmarks. (Pl. H.) Average size of 100 eggs, $1\cdot35 \times \cdot97$ in. [$34\cdot5 \times 24\cdot8$ mm.]. The nests are generally to be met with in colonies of varying size, and are found in shallow lakes or marshes overgrown with water-lilies, rushes, and other forms of aquatic vegetation. The usual breeding time is from about May 20 to the first week of June in Central Europe, but in South Spain eggs may be found by the middle of May. Naumann states that both sexes incubate in turn, but the hen alone at night, and that the incubation period lasts 14-16 days.[1] Only one brood is normally reared in the season, but if the first clutch is taken, another is deposited soon afterwards. [F. C. R. J.]

5. **Food.**—Chiefly aquatic insects and their larvæ, also land insects, earth-worms, and occasionally, according to Naumann, little frogs, spawn, and small fish. [F. B. K.]

COMMON-TERN [*Sterna hirúndo* Linnæus; *Sterna fluviátilis* Naumann. Sea-swallow, sparling, skrike (generic), kip (Kent). French, *hirondelle de mer, Pierre-Garin* ; German, *Fluss-Seeschwalbe* ; Italian, *rondine di mare*].

1. **Description.**—The common-tern is to be distinguished by the orange-red colour of the beak, which is black-tipped, the pale vinaceous grey under parts, the broad band of dark grey which runs along the inner side of the white shaft of the outer primaries (see Arctic-Tern, p. 61), and the length of the tarso-metatarsus, which is longer than the middle toe minus the claw. (See also p. 76.) The sexes are alike, and there is an incomplete winter, as distinct from the summer,

---

[1] See, however, notes on the incubation period under sandwich and little-tern.

dress. (Pl. 99.) Length 14·25 in. [361·94 mm.]. The forehead, crown, and nape are black, the interscapulars, scapulars, and wings are of a dark pearl-grey, the rump and tail feathers white, the latter tinged with grey along the outer webs, darkest on the "streamers." The outermost primary has the outer web dark slate colour, the inner with a band of dark grey along the shaft, the rest of the web being white except towards the tip, which is dark ash-grey; inner primaries paler grey with white "wedges." The secondaries are tipped with white. The throat is white, the rest of the under parts pale lavender-grey. The beak is orange-red, dusky at the tip; the legs and toes coral-red. After the autumn moult the fore-part of the crown becomes largely intermixed with white, and the under parts are almost pure white. In the juvenile plumage the pearl-grey of the upper parts is darker than in the adult, and obscured by broad terminal fringes of deep buff and subterminal bars of slaty black. The forehead and fore-part of the crown are buff coloured, and the hind-crown and nape dull black; the lores are also black. The inner secondaries are tipped white, and have subterminal loops of dark brown. Primaries much darker than in the adult, and with subterminal bars of dull slate-black, and a strongly marked white line along the free edge of the inner web. Tail much darker grey than adult. In some individuals the mantle is marked with wavy lines of black and brown, and the hindmost scapulars and inner secondaries have conspicuous loops of dull slate-black alternating with loops of white. In such birds the white of the forehead is thickly spotted with dark grey obscuring the white. The inner wing-coverts are powdered with black and buff. In other individuals the black is conspicuous by its absence and the pearl-grey clouded with brown, as also are the forehead and fore-part of the crown, but the lores, hind-crown, and nape are dull black. After the autumn moult the only sign of immaturity is a band of dark grey across the upper wing-coverts. The young in down (protoptyle) are of a pale smoke-grey, tinged with buff, except the cheeks, which are pure grey. This ground-colour is relieved by black markings, which on the back may form two indistinct longitudinal stripes. In some chicks the ground-colour of the upper parts is much deeper. A live specimen noted at Romney was dusky buff with an unusual number of black irregular markings (F. B. Kirkman). The same varia-tion has been noted at Ravenglass (H. W. Robinson, *British Birds*, iii. 169). Both at the latter colony (*loc. cit.*) and at Muskeget, off the New England coast, the chicks have been noted to vary in the coloration of the legs, some having them reddish, others flesh-coloured or yellowish (G. H. Mackay, *Auk*, xiii. 52). The

examination of a large number of live chicks shows that they vary in the coloration of the chin, most having black chins with a small white patch at the base of
the lower mandible, some no black. The black grows lighter with age, becoming
grey and then white (T. A. Coward, *in litt.*). The under parts are white.
[w. p. p.]

2. **Distribution.**—In the British Isles this species is widely distributed
where the coast furnishes suitable breeding ground, but in Scotland and the north
of England, as well as in Ireland, its range overlaps that of the Arctic-tern, which
outnumbers it in Northern Scotland. Both species are also found breeding on the
Irish coasts, but the common-tern is the more numerous species in the south of Great
Britain, though absent from some districts, such as the Devon and Somerset coasts.
As both the common and Arctic terns are somewhat capricious in their choice of
breeding-places, and are apt to shift their ground from year to year, it is difficult
to define their respective ranges accurately. In the Outer Hebrides the former
is by no means common, but has long been known to breed in the Orkneys, and
since 1901 has nested also in the Shetlands in some numbers. South of the Farnes
on the east side of England, and Anglesey on the west, it practically replaces the
Arctic-tern altogether, though apparently both species formerly bred on the Scillies.
In Ireland it is less numerous than the Arctic-tern, and shows a greater tendency
to breed on fresh-water lakes. On the Continent its range extends over the coasts,
lakes, and larger rivers from the North Cape of Norway and the White Sea, south
to the Mediterranean, Black and Caspian Seas. It is also found in the Canaries and
Azores, and along the coast of North Africa, while in Asia it occurs in the Ob and
Yenesei valleys, and south to Transcaspia and Mesopotamia, but in Central Asia
and the Lena valley it is replaced by other forms. In North America it is plentiful
on the Atlantic side from Labrador south to Texas. The winter range in the Old
World extends over the Continent of Africa south to Cape Colony, and in Asia to
India, Ceylon, the Malay Peninsula, etc. ; while in America it has been recorded
exceptionally as far south as Bahia, Brazil, but generally reaches the northern States
of South America (Venezuela, Guiana, etc.). [F. C. R. J.]

3. **Migration.**—A summer visitor. The earliest arrivals on the Kentish
coast occur in the second week in April, and the birds continue passing northward
till the middle of May (cf. Ticehurst, *Birds of Kent*, p. 501). They were reported at
Blakeney, Norfolk, in 1910, on April 28 (cf. *Zoologist*, 1911, p. 174), and at Ravenglass about April 25. They seldom appear on the Welsh coast until the last week
in April (cf. Forrest, *Fauna of N. Wales*, 1907, p. 374), and Ireland and Scotland

are seldom reached until May. The southward passage lasts from August to October, during which period they are seen not only along the coast, but inland (cf. O. V. Aplin, B*irds of Oxfordshire*, p. 166). Few remain till October, and some have reached the Spanish and Portuguese coasts by September (see below). The species is scheduled in the B. *O. C. Migration Reports*, but very little definite information has as yet been collected. As regards habits, the species is said to appear at its summer haunts in the morning, having arrived during the night, the numbers increasing nightly for ten or twelve nights (*Migration Report*, 1883, p. 45). Several interesting records of marked common-terns recovered, exist : (1) marked at Ravenglass, Cumberland, July 30, 1909, recovered at Espiña, Galicia, Spain, September 21, 1909 ; (2) marked at Ravenglass July 23, 1910, recovered 25 miles south of Oporto, Portugal, about September 12, 1910 ; (3) marked at the Hook of Holland July 16, 1910, recovered at Boulogne, France, September 1, 1910 (cf. Witherby, B*ritish Birds*, iii. p. 219, and iv. p. 179). [A. L. T.]

4. **Nest and Eggs.**—The nesting-sites vary extraordinarily : off the west of Scotland one may find large colonies breeding on the bare, water-worn rocks ; in East Anglia and other parts the eggs are laid on the great shingle-beds close to the sea ; in Holland the nests are sometimes on short meadow grass or in sandy mud-flats in the polders, and at other times floating on the surface of fresh-water meers inland. In some cases there is no lining whatever, a natural cavity in rock being used or a hollow in sand : at other times dry bents, bits of dead reed, and other vegetable matter are carelessly arranged. (Pl. XLIII.) The eggs are 2 to 3 in number, generally the latter ; but I have several times seen 4 or 5 together, pro-bably laid by more than one bird. The ground-colour varies from dull greyish to ochreous and brownish, blotched and spotted, sometimes boldly and richly, but more often rather sparingly with dark blackish brown and underlying ashy grey. Some eggs when first laid have a very beautiful blue ground, which unfortunately soon fades, and a distinctly red variety also occurs on rare occasions. (Pl. H.) Average size of 108 eggs, 1·61 × 1·19 in. [40· × 30·2 mm.]. Incubation commences as soon as the first egg is laid, and apparently lasts 20-21 days from the date of the full clutch. This is confirmed by W. Evans's observations on eggs hatched in incubators, which hatched on the twenty-second and twenty-third days. Naumann states that both sexes incubate, but chiefly the hen, who also broods during the night. In fine sunny weather the eggs are left uncovered for long periods. Saunders states that exceptionally eggs have been found by May 15 ; but this is very unusual, and as a rule they are laid about the beginning of June,

but in the north often not till the middle of the month.   Only one brood is reared
in the season.   [F. C. R. J.]

5. **Food.**—Chiefly small fish, aquatic insects, and small crustaceans.   In
twenty-one stomach contents examined by Rörig were found fish remains, small
gnats (Mücken), flies, dragon-flies, and earwigs (Naumann, *Vögel Mitteleuropas*, xi.
133).   Earthworms are said to be sometimes taken.   The young are fed by both
parents, chiefly on small fish.   At Ravenglass these were found to be chiefly young
herrings, but "many small whiting were found, and also a few young codling,
lumpsuckers, and long rough dabs, and, although the colony was bounded on one
side by a river famous for its *Salmonidæ*, no trace of the young of these fish was
found at all on the ground" (H. W. Robinson and F. W. Smalley, *British Birds*, iii.
169).   [F. B. K.]

ARCTIC-TERN [*Sterna paradisæa* Brünnich ;[1] *Sterna macrúra* Naumann.
Sea-swallow, sparling, skrike (generic), tarrock, piccatarries (Shetlands).
French, *hirondelle de mer arctique* ; German, *Küsten-Seeschwalbe* ; Italian,
rondine di mare coda lunga].

1. **Description.**—The Arctic-tern may readily be distinguished from the
common-tern by the uniform blood-red colour of the beak,
the dark grey of the under parts, longer tail streamers,
the shorter tarso-metatarsus, which does not exceed the
length of the middle toe minus the claw, and by the
narrower width of the grey band along the shafts of the
outer primaries.   (See accompanying figure, and also

1. COMMON-TERN'S PRIMARY.

p. 76.)   The sexes are alike, and there is a more or less decided seasonal change
of coloration.   (Pl. 100.)   Length 14·5 in. [368·30 mm.].
The adult, in summer, has the top of the head, from the
base of the beak backwards to the nape, black, the rest
of the upper parts dark pearl-grey, fading into white on
the lower rump and upper tail-coverts.   The tail is white,
but the feathers have a faint grey tinge on the outer

2. ARCTIC-TERN'S PRIMARY.

webs, and dark grey outer webs to the outermost elongated feathers or "streamers."
The outer web of the outermost primary is blackish, and along the inner side

---

[1] Brünnich's name as given above was published in 1764, two years before the issue of the
12th edition of Linnæus.   It was therefore ignored by Saunders, who made the 12th edition
of the *Syst. Nat.* his starting-point, but must now be adopted in accordance with the rules of
the Fifth International Zoological Congress. [F. C. R. J.]

of the shaft runs a band of grey, much narrower than in the common-tern, leaving the rest of the web white save near the tip, where it is grey. The under parts are of a dark grey save the throat and under tail-coverts, which are white. After the autumn moult the black of the fore-part of the crown becomes intermixed with white, and the under parts are paler. The fledgling resembles the young of the common-tern at the same age, and can be distinguished therefrom only by the dark grey colour of the outer tail feathers. After the autumn moult the young resemble the adults in winter, having the forehead and crown nearly white, but a dark grey band on the upper wing-coverts, more grey on the outer webs of the tail feathers, and the under parts white. The young in down, like those of the common species, show variation in their coloration. The ground-colour of the upper parts is (1) either creamy white, white, or greyish, or (2) shades of buff. In both the markings are black, and the under parts white with a greyish tint about the flanks and vent. The coloration of the throat varies in the same type, and no doubt changes with growth. It is usually black or dusky. The legs and beak vary from red to flesh colour. The beak has a dusky tip, which is lost later (N. F. Ticehurst in *British Birds*, iii. 200; F. B. Kirkman, *in litt.* Observations based on live specimens). [w. p. p.]

2. **Distribution.**—In Great Britain this is a more northerly breeding bird than the common-tern, and does not nest south of the Farnes on the east coast of England, nor along our southern shores. There was, however, formerly a large colony on the Scillies, but at the present time the southern limit of this species on the Welsh coast appears to be the Skerries and the Anglesey coast. On the western coast and islands of Scotland it is very plentiful locally, and is also common in the Shetlands and Orkneys, while it is the predominant species in Ireland, many large colonies existing on marine islands. On the whole it is less of a fresh-water haunting species than the common-tern, but colonies are known on Loughs Corrib, Mask, and Carra in Ireland, as well as in the Orkneys. Outside the British Isles it breeds in the Færoes, Iceland, Franz Josef Land, Spitzbergen, Scandinavia from Bergen northward along the Norwegian coast to Lapland; Lake Onega, the lower Dwina and Petchora, in the Baltic on the coasts of Finland, Esthonia, and Sweden; Denmark, the N. Frisian Isles, while recently colonies have been discovered in Holland. In Asia it is found along the N. Siberian coast and the main river valleys east to the Kolyma Delta, Chukchiland, and the Anadyr and Commander Isles, and, according to Radde, also in Baikalia, while in N. America it breeds in Greenland to over 82° N., the shores of the Arctic Ocean, and the Atlantic coast south to Massachusetts. The winter range of this species in Africa extends to Cape Colony and even as far south

PLATE XLIII

Photo by F. B. Kirkman

Well-lined common-tern's nest in marram grass (Walney)

Photo by F. B. Kirkman

Arctic-tern sitting on its nest in bare sand (Walney)

Photo by J. Atkinson

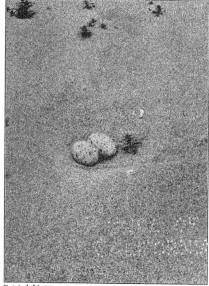

Photo by J. Atkinson

as 74° 1′ lat. (*Ibis*, 1907, p. 345),[1] while specimens have been obtained in lat. 66° S. beyond New Zealand, and off the coasts of Peru, Chile, and Brazil in S. America. [F. C. R. J.]

3. **Migration.**—A summer visitor. In spring migrants are seen passing north along the coast of Kent from mid-April onwards (cf. Ticehurst, *Birds of Kent*, p. 503). A few early arrivals are generally recorded from Lancashire in the last week of April (cf. B. O. C. *Migration Reports*). In 1911 the first arrivals reached the Farnes on April 21 (H. A. Paynter). The usual time of arrival for the majority of the birds is the end of April and early May; in Scotland later. The date given for the Outer Hebrides is mid-May (Harvie-Brown and Buckley, *Fauna of Outer Hebrides*), and for the Orkneys May 14-18 (*Annals Scot. Nat. Hist.*, 1904, p. 23). The departure extends from August till October, but precise records are lacking. A passage along our coasts of Arctic-terns summering farther north may be presumed to exist, but there is no information on this head either. An Arctic-tern marked as a chick on the Farne Islands on July 17, 1909, was caught at the Barns Ness Light, East Lothian, on August 23, 1909 (cf. *Country Life*, October 16, 1909, p. 543). A young Arctic-tern marked on Sylt (North Frisian Islands) on June 27, 1911, was picked up near Pontefract, Yorkshire, on August 20, 1911 (cf. Weigold, quoted by Witherby, *British Birds*, vol. v. p. 130). [A. L. T.]

4. **Nest and Eggs.**—The breeding colonies of this species are sometimes to be found on sand-banks or shingle-beds, at other times on bare rocks, and occasionally near the shores of lakes on ground covered with a scanty vegetation. In the Orkneys it has been found breeding on grass and in cornfields (*Annals of Scot. Nat. Hist.*, 1892, p. 78). Sometimes no nest whatever is made, the eggs being laid in natural hollows of the lichen-covered rocks or in scratchings in the sand. At other times a little grass, blossoms of thrift, a few bents or bits of seaweed and wrack are used to line the nest hollow. Some Shetland nests are described as thick structures of broad herbage : one built of seaweed was 8 inches across. (See also article by F. B. Kirkman in *British Birds*, ii. p. 78, on variation in type of nest, and photos.) (Pl. XLIII.) The eggs vary in number from normally 2 to occasionally 3 ; in many localities 2 are almost universally found. Thus T. E. Buckley states that in the Orkneys the clutch consists of 2 only, while on the east Sutherland coast 3 are frequently laid. Clutches of 4 eggs are the produce of more than one hen. They

[1] Richenow (*Ornitholog. Monatsber.* 1907, p. 107) holds the opinion that these birds are *Sterna antistropha*, a closely-allied species, or more probably sub-specific form, whose breeding grounds are unknown, but apparently exist somewhere in the southern hemisphere. See also *Deutsche Südpolar-expedition*, 1901-1903. *Vögel des Weltmeeres*, p. 463 (A. Reichenow). [F. C. R. J.]

are more variable in colour and markings than those of the common-tern, the ground-colour ranging from deep rich red-brown through all shades to pale stone colour and light blue. On the average the markings are also richer and bolder, and the average size is rather less, but some undoubted eggs of the common-tern are quite as small and richly marked as any Arctic-tern's eggs. (Pl. H.) Average size of 116 eggs, 1·6 × 1·16 in. [40·8 × 29·5 mm.]. Incubation lasts 20 days according to Mr. F. G. Paynter, and Naumann states that both sexes brood in turn but the hen alone during the night, though during bright and sunny weather the eggs are left exposed for long periods. Naumann's statement that the incubation period in the case of this species (and also the common-tern) lasts only for 15-16 days cannot be relied upon. The breeding season is slightly later than that of the common-tern, and in the British Isles eggs are rarely to be found before the beginning of June or the end of May at the earliest, but on the west coast of Jylland I found birds just beginning to lay on May 12, an exceptionally early date. Only one brood is reared during the season. [F. C. R. J.]

5. **Food.**—Small fish, small floating crustaceans, insects and their larvæ, earthworms. Mr. H. C. Hart, in his account of the "Ornithology of the British Polar Expedition, 1875-76" (*Zool.*, 1880, p. 206), states that at Musk Ox Bay, at 81° 45′ N. lat., their "chief food was green caterpillars" (*Argynnis chariclea* and the larvæ of *Tipula arctica*). The young are fed by both sexes, chiefly on small fish, and sometimes on earth-worms and insects. An instance of the species feeding its young on crane-flies, greendrakes, and May-flies is given in *British Birds*, iii. 91, by Mr. Wm. C. Wright—and on earth-worms by Mr. J. Tomison in the *Annals of Scottish Natural History*, 1904, p. 23. [F. B. K.]

ROSEATE-TERN [*Sterna dougálli* Montagu. French, *sterne de Dougall*; German, *Dougalls-Seeschwalbe*; Italian, *rondine di mare del MacDougall*].

1. **Description.**—The roseate-tern bears a close likeness to the common-tern, from which, however, it may be distinguished by the long and slender black beak, the relatively shorter wings, wherein the inner webs of the primaries are white to the tips, the relatively longer and more deeply forked tail, and the rosy-tinted breast. There is a scarcely perceptible seasonal change of coloration, and the sexes are alike. (Pl. 101.) Length 15·5 in. [393·69 mm.]. The adult, in summer dress, has the top of the head and nape black, the pearl-grey of the upper parts paler than in the Arctic and common terns. The white inner margins of the primaries extend to

the tips of the feathers and round to the outer web for a short distance, and the white of the under parts is suffused with a delicate rose-pink. The beak is black, with a small patch of orange at the base, and the legs and toes are red. In winter the forehead is more or less white, and the rose-pink of the breast is only faintly traceable. The fledglings may be distinguished from those of the common and Arctic terns by the white on the inner webs of the primaries, which indeed is distinctive at all stages and in both sexes. Skins of downy young, examined at the British Museum, have the upper parts buff, mottled with grey and white, and the under parts dull white. Live specimens examined by Mr. G. H. Mackay (*Auk*, xiii. 47) were grey, white, and black on the upper parts, and whitish under. Their legs and feet were black. The beak pinkish flesh colour, with black tip. This points to dimorphism. [w. p. p.]

2. **Distribution.**—In the British Isles this is now an extremely local species, but one very populous colony still exists within our limits, although the indiscriminate protection afforded there to both the roseate and the common tern may prove prejudicial to the rarer species. In England, although carefully protected on the Farnes since 1888, only two pairs were breeding in 1899, and in 1900 only one pair (see *Zoologist*, 1902, p. 52). It has been seen off the Norfolk and Suffolk coasts in the breeding season, but has not been proved to breed there except casually, and formerly a colony existed on the Scillies, which has now disappeared. There is, however, a colony on the Skerries, and another much larger colony also exists in North Wales, where the numbers are well maintained. The Walney and Foulney Islands are now deserted; and in Scotland it no longer breeds in the Firth of Clyde, but Mr. O. A. J. Lee found two pairs breeding on the Culbin Sands in Moray in 1887. In Ireland the breeding colonies off the Dublin and Down coasts have long ceased to exist, but there is a possibility that it still nests among the other terns on the west coast, although proof is still wanting. On the Continent it breeds on the coast of Brittany, and is said to nest in Holland, but this statement requires confirmation. It undoubtedly breeds in South Tunisia, possibly in Madeira, and in large numbers on islands off the Somali coast, between the Seychelles and Mauritius, and in the Andaman Isles, where it is said to be represented by a small race. Another local form also breeds in Eastern Asia, Japan, Fokien, etc., and large colonies exist on Houtmans Abrolhos, as well as in the Moluccas and New Caledonia, while in America breeding colonies exist from Massachusetts and New York to the West Indies, Central America, and Venezuela. In the case of a bird of cosmopolitan distribution it is difficult to ascertain

the migration limits, but Saunders obtained specimens from Cape Colony, and in S. America it has been obtained at Bahia in Brazil. [F. C. R. J.]

3. **Migration.**—A summer visitor which is said to arrive at the very end of April and to leave as soon as the young can fly (cf. Saunders, *Ill. Man. B. B.*, 2nd ed., 1899, p. 646). But the relative scarcity of the species has made accurate migrational study very difficult. [A. L. T.]

4. **Nest and Eggs.**—In most cases the breeding colonies are found on bold rocky islets, but in some cases the eggs are recorded as having been found among sand and shingle. Most of those which I have personally examined were placed in recesses or ledges of rocks, or in miniature caves, sometimes sheltered by the coarse vegetation growing on the rocks. In only a few cases was any nest material used, the eggs being laid on the bare rock or shingly ground. The full clutch consists of 2 eggs, but in some cases only a single egg is laid. Quite exceptional cases of clutches of 3 have occasionally been recorded. Normally they are readily distinguishable from those of the Arctic and common terns, though occasionally eggs of the latter species are indistinguishable. The ground-colour varies but little, being almost always light stone colour, occasionally suffused with brownish. The markings consist of small spots and speckles of chocolate to sepia, with underlying blotches and spots of ashy grey, and in some cases the dark markings form a very distinct zone round the big end of the egg. Not a single egg out of a large number examined showed the varied ground-colours prevalent in a colony of Arctic-terns. (Pl. H.) They are also large for the size of the bird, and elongated in shape: average size of 37 authentic eggs, 1·74 × 1·16 in. [44·4 × 29·5 mm.]. Accurate observations on the length of the incubation period and the share of the sexes in the work are still wanting, but probably in these respects they differ little from their congeners. The breeding season is rather late, and the eggs are not as a rule laid till June, generally about the first week in that month. Only one brood is reared during the season. [F. C. R. J.]

5. **Food.**—Chiefly fish and floating marine organisms. Information is scanty. [F. B. K.]

SANDWICH-TERN [*Sterna sandvicénsis* Latham ; *Sterna cantíaca* Gmelin. Sparling, kek-swallow, skrike, pearl-gull (Devon), schreecher (Kent). French, *hirondelle de mer caugek* ; German, *Brand-Seeschwalbe* ; Italian, *beccapesci*].

1. **Description.**—The sandwich-tern is readily distinguished by its stouter

build, black beak and legs, and the long pointed feathers of the nape, which form a kind of crest. (See also p. 101.) The sexes are alike, and the only seasonal change of plumage is confined to the black area of the head. (Pl. 102.) Length 16·5 in. [419·10 mm.]. In summer the forehead, crown, and nape are black, the nape feathers being long and pointed, forming an incipient crest; the rest of the upper parts pearl-grey save the rump and tail, which are white. The under parts are white, tinged with a delicate salmon-pink. The beak is black with a semitransparent yellowish tip, and the legs and toes are black. After the autumn moult the fore-part of the crown is white, and the nape is mottled with white. The juvenile plumage of the crown is barred with black and white, and the rest of the upper parts are variegated by submarginal dusky loops. A dusky band extends along the wing formed by the marginal and minor coverts, and the tail feathers have a subterminal bar of black. The under parts are white. The downy young are buff coloured, irregularly mottled with grey and black, but around the eye are a number of irregular loops of black, and there is a more or less distinct black line over the scapular region, bounded on either side by a band of buffish white. (See also p. 105.) [w. p. p.]

2. **Distribution.**—In the British Isles the colonies of this species are not numerous. It no longer breeds regularly in the Scillies, Kent, or Norfolk, though attempts have been made to do so of late years. At the Farnes, however, as well as at Ravenglass in Cumberland, there are large colonies, although Walney is now deserted. In Scotland it has been known to nest in some numbers near the mouth of the Findhorn, and was found breeding on N. Ronaldshay in the Orkneys in 1893, while it formerly bred on Loch Lomond, and does so at the present time in Kircud-bright. In Ireland there are now several thriving colonies in Co. Mayo (near Ballina and on Lough Conn), one in Fermanagh (on Lough Erne), and two small colonies on marine islands in Down. On the Continent there are large colonies in Jylland, and in the Cattegat, Denmark, and in the N. Frisian Isles, as well as off the W. German and Dutch coasts, and in the Channel Islands. A few breed in Spain, as well as in the Canaries, and in the Mediterranean it nests commonly in Tunisia, and is said to breed in Sardinia, Sicily, and on the coast of Italy, as well as in some numbers in the Black and Caspian Seas. In America it is replaced by a closely allied form, which breeds from the New England coast south to Honduras. The winter range of this species in Africa extends south to Cape Colony and Natal: in Asia it ranges to the Red Sea, Persian Gulf, and the Sind coast, while the American form has been recorded from Colombia. [F. C. R. J.]

3. **Migration.**—A summer visitor. The earliest arrivals are seen on our shores towards the end of March. The dates of first arrivals recorded by the watcher at Ravenglass for 1907, 1908, 1909 were respectively 28th, 27th, and 29th March. In Scotland the first records are generally for the beginning of May. In Ireland a March arrival is rather the rule than the exception. For a long list of records see Ussher and Warren, *Birds of Ireland*, p. 317. The dates of departure are not well known. By the 23rd of July in 1909 the birds had all left Ravenglass (cf. *British Birds*, iii. 170). On August 20 of the same year, a young bird of the year ringed at Ravenglass in June was captured thirty to forty miles north of that place, thus tending to show that the southward movement is not general in August (*op. cit.*, p. 181). On the Yorkshire coast birds are seen moving south in August, but small parties remain till September and even October (Nelson, *Birds of Yorks.*, p. 652). They reach the Kent coast about the end of September, and stay only a few days before going south (cf. Ticehurst, B. *of Kent*, p. 499). [A. L. T.]

4. **Nest and Eggs.**—This species breeds in colonies on sandy flats, often with a scanty growth of plants, and in many cases the nests are mere hollows scratched in the sand, though occasionally fairly substantial nests of bents and dead vegetable matter cast up by the sea may be met with. Its nests may also be found on shingle, on mudbanks, and in sea-campion (see p. 102). (Pl. XLII.) The eggs as a rule are 2 in number, and clutches of 3 are quite exceptional in most colonies, though occasionally a small group may contain two or three such nests. On one occasion 4 eggs were found in a nest, but it is very doubtful whether they were all laid by one bird. They are remarkably handsome and show considerable variation, the ground-colour ranging from greyish or creamy white to warm stone colour, sometimes suffused with brown. The markings consist of spóts and blotches of dark sepia or rich red-brown exceptionally, varying in size and sometimes forming a zone, while in some eggs they are almost absent, or replaced by fine specks. The underlying markings are ashy bluish or pale brownish, and in some eggs are very prominent, consisting of large blotches. (Pl. H.) Average size of 100 eggs, 2·03×1·42 in. [51·7×36·1 mm.]. Incubation begins as soon as the first egg is laid, and, according to Mr. F. G. Paynter's observations, lasts for 20 days (Naumann gives three weeks as the period). Both sexes share in the work (see p. 103). Like the other terns, in fine weather they leave their eggs exposed to the sun for long periods. The breeding season is decidedly earlier than that of the other British species, and eggs may occasionally be found at the end of the first

week in May, while laying is general about the middle of the month. Where the first clutches have been taken, fresh eggs may of course be obtained later, but only a single brood is reared during the season, and by about the middle of July the breeding-places are entirely deserted. [F. C. R. J.]

5. **Food.**—Chiefly small fish. Also " small shell-fish." The stomach of one bird was found full of the " comminuted remains of a bivalve shell " (Macpherson, *Fauna of Lakeland*, p. 411). The young are fed on small fish, young whiting being quoted as the principal diet in one colony, Ravenglass (H. W. Robinson and F. W. Smalley, *British Birds*, iii. 169). [F. B. K.]

LITTLE-TERN [*Sterna minuta* Linnæus. Little-kip, skerrek. French, *petite hirondelle de mer* ; German, *Zwerg-Seeschwalbe* ; Italian, *fraticello*].

1. **Description.**—Apart from its small size the lesser-tern can at once be distinguished by the broad white forehead and broad black stripe from the beak to the eye. The sexes are alike, and there is a slight seasonal change of plumage. (Pl. 98.) Length 9·5 in. [241·30 mm.]. There is a broad band of white on the forehead, but the rest of the crown and nape are black. The upper parts are of a pearl-grey colour, save the rump and upper tail-coverts and tail feathers, which are white, and the primary major coverts, which, like the first three primaries, are blackish along the outer web and the inner side of the shaft; the remainder of the primaries are pearl-grey, with white margins to the innermost. The outer secondaries have the outer webs dusky grey, the rest white ; the inner are pearl-grey, like the back. The sides of the head and under parts are of a silky white. Beak gamboge-yellow, tipped with black, legs and feet orange-yellow. After the autumn moult the white on the forehead broadens somewhat, and the outer primaries are darker towards the tip. In the juvenile plumage the feathers of the upper parts have dusky horse-shoe, or arrow-headed, subterminal bars, and are tinged with sandy buff, but the rump, tail-coverts, and tail feathers have a wash of pearl-grey. The wing-coverts are variegated like the back, save the marginal coverts, which are dusky, forming a dark band across the wing. The primaries are of a dark slate edged with white, and conspicuously different from those of the adult. The forehead is of a dull white, and the crown and nape dusky black. The young in down is of a light sandy buff or sandy grey, the nape spotted with black, and with three fairly distinct longitudinal stripes of black along the back. Under parts white. [W. P. P.]

2. **Distribution.**—Unlike the Arctic, roseate, and common terns, the little-

tern nests only on sandy and shingly beaches, and not on rocky islands. Scattered colonies may be found along our eastern and southern coasts from Teesmouth and the Spurn southward, but it no longer breeds in Northumberland. On the west side colonies exist in Cornwall, Wales, the Isle of Man, and the Cumbrian coasts, but in Scotland it is local, though known to breed in S.-W. Scotland, Tiree, and the Outer Hebrides, as well as in the Orkneys, and on the east side reported as common and increasing from Tay to Dee by Mr. Harvie-Brown. The Haddingtonshire colony, however, is now deserted. In Ireland, though much less numerous than the common and Arctic terns, there are many colonies on the coasts of Leinster, Ulster, and Connaught. On the Continent it breeds in Denmark, in Sweden on the coasts of Skåne, Öland, and Gotland, in Russia only south of Esthonia and the Perm government. South of these limits its range extends to the Mediterranean; and it is known to breed in the Canaries and Madeira, along the North African coast, in the Balearic Isles, Sicily, Greece, Asia Minor, and Cyprus (occasionally) to the Black and Caspian Seas. On the Syrian coast and Lower Egypt it is replaced by the black-shafted form, but in Asia it occurs in the S. Tobolsk and Turgai governments and at Barnaul and on the middle Irtysh, while it is also said to have bred in Northern India, but on the Indian coast it is replaced by the black-shafted form. Other allied species replace it in North and South America. On migration it is found in Asia as far as Burma, the Malay Peninsula, and even in Java, while in Africa it certainly occurs in Gambia, Nigeria, and the Gold Coast, and is said to have once been obtained in Cape Colony. [F. C. R. J.]

3. **Migration.**—A summer visitor. In Kent a few birds have been seen as early as April 9, and a fair number generally appear about the 20th, but the majority of these pass on. The majority of the Kentish breeding birds arrive towards the end of April and the beginning of May (cf. Ticehurst, B. *of Kent*, p. 504). In 1910, little-terns arrived at the Blakeney colony (Norfolk) on April 26 (cf. Q. E. Gurney, *Zoologist*, 1911, p. 174). Their usual time for appearing at Spurn Point is the end of April (cf. O. Grabham, *British Birds*, ii. 319), as also in N. Wales (cf. Forrest, *Fauna of N. Wales*, p. 376); but in Ireland they are rather later (Ussher and Warren, *Birds of Ireland*, p. 326). They frequently do not appear on the Scottish coasts till the second week of May. The first arrivals have appeared in recent years between the 5th and 20th May. Most of the birds leave their breeding-grounds in August. The southward migration takes place chiefly in September, there being few birds seen about our coasts after the end of that month. [A. L. T.]

4. **Nest and Eggs.**—The breeding-place is generally only a hollow in the sand or among the shingle on the beach. In the latter position the eggs are very difficult to see, though easily found on sand. In many cases the hollow is lined with fragments of cockle-shells or pebbles. (Pl. XLIII.) The eggs are 2 or 3 in number, much smaller than those of our other British breeding terns. The ground-colour varies from light stone to ochreous, suffused with brown, blotched and spotted with dark brown, which in some cases seems to have "run," and ashy grey underlying blotches. (Pl. H.) Average size of 123 eggs, $1.25 \times .92$ in. [31.8×23.5 mm.]. According to Naumann both sexes incubate, but chiefly the hen; while in fine weather the eggs are left uncovered for long periods. He also gives the incubation period as 14-15 days, but this is probably incorrect, as Mr. T. Hepburn's and Mr. Oxley Grabham's observations tend to fix the period at from 17-20 days (*Zoologist*, 1904, p. 173; *British Birds*, ii. 319), and Naumann's estimates in the case of the other terns are also much under the mark. The breeding season begins in England about the last week of May or early in June, but occasionally eggs may be found as early as May 17, and only a single brood is reared during the season. [F. C. R. J.]

5. **Food.**—Chiefly small fish and crustaceans. Dr. Patten states that he has "seen little-terns, especially immature birds, foraging with turnstones under rotting seaweed for sand-hoppers" (*Aquatic Birds*, p. 398). Aquatic insects and their larvæ (Naumann, *Vögel Mitteleuropas*, xi. 124). The young are fed by both sexes on small fish. At Spurn Point they were found to be fed "largely on very small plaice about the size of a penny, sand-eels, sprats, etc." (O. Grabham, *British Birds*, ii. 320). [F. B. K.]

The following species and subspecies are described in the supplementary chapter on "Rare Birds" :—

Whitewinged black-tern, *Hydrochelidon leucoptera* (Temminck).

Whiskered-tern, *Hydrochelidon leucopareia* (Temminck). [*Hydrochelidon hybrida* (Pallas).]

Gullbilled-tern, *Gelochelidon nilotica* (Gmelin). [*Sterna anglica* Montagu.]

Caspian-tern, *Sterna caspia* Pallas.

Sooty-tern, *Sterna fuliginosa* Gmelin.

[Lesser sooty-tern, *Sterna anæstheta* Scopoli].

[Noddy, *Anous stolidus* (Linnæus)].

THE BLACK-TERN

[F. B. KIRKMAN]

If we set aside the noddy (*Anous stolidus*), which is of very doubt-ful occurrence in the British Isles,[1] and the rare gullbilled-tern (*Gelochelidon nilotica*), our Terns fall into two groups, which may, roughly speaking, be distinguished as the Marsh-Terns (*Hydro-chelidon*) and the Sea-Terns (*Sterna*). The latter frequent chiefly the sea and estuaries, and their principal food consists of fish; the former are birds of marsh and fen, and feed chiefly on aquatic insects. The chief British representative of *Hydrochelidon* is the black-tern, its congeners, the white-winged black-tern (*H. leucoptera*) and the whiskered-tern (*H. hybrida*) being very rare exceptional visitors. The black-tern is now also a comparatively rare visitor. At the beginning of the last century it bred in hundreds in the Fens, the Norfolk Broads, and possibly elsewhere, but it has been driven out of its ancient haunts partly by cultivation and partly by persecution. It visits them still in the spring on its way north to other breeding-grounds in Denmark and the Baltic countries. In autumn, again, it passes them with its young on its way to its African winter quarters.

For an account of its habits we have to depend chiefly on foreign sources, and these are very scanty, the information being of the usual type supplied in the standard works. Though rare with us, the species is very numerous as a summer visitor in certain parts of Europe, for instance in Hungary and Holland. It arrives at its breeding quarters in May. When migrating to or from these it flies at a great height, from which it descends from time to time when feeding quarters present themselves. The migration appears to be performed in a leisurely manner, and the birds may pause for some days at a spot which provides abundance of nourishment. Thousands migrate along the Danube, following each bend in the river, feeding on their way,

---

[1] See R. J. Ussher, *List of Birds*, and T. A. Coward, *Fauna of Cheshire*, i. 93.

and flying low, contrary to their usual custom when migrating over-
land or oversea.[1]

The species breeds in colonies, large or small. Solitary pairs are
very exceptional. The nests are built in marsh or fen-land, on the
ground, on floating masses of water-plants, sometimes on standing
bent and matted reeds, occasionally on water-lily leaves, when these
are close and strong enough to bear the nest and its occupants.
In Southern Spain Mr. W. Farren found the nests floating in the
water insecurely anchored by an occasional thin rush. The disturbance of the water caused by his wading in made them bob up and
down; one, indeed, was shipwrecked. The material of which they
were constructed was scanty, and in the surface scum of the water
difficult to see. They were in groups of three or four, the groups
being usually a hundred yards or so apart. .

Whether both sexes build is not recorded. The material used is
chiefly water-plants, and these are often picked from the surface of
the water when the bird is on the wing. The eggs are incubated,
according to Naumann, by both birds, and the two to three eggs are
hatched after about a fortnight, which is almost certainly an under-
estimate. According to the same authority the young remain in the
nest for two weeks, and are fed by the parents with insects. They
continue to be fed both on the ground and in the air for some time
after leaving the nest. The parents show great excitement when the
nesting-ground is approached. In Holland Mr. Jourdain saw one
strike a boy on the head and pursue him for quite a distance,
threatening to strike again, though at the time it had not even laid.
Old and young depart, for the most part, in August. Of their intimate
home-life nothing seems to be known beyond these few facts.

Those observers who have noted the bird's habits during its short
visits to us—most are content merely to record its occurrence at this
or that place—have provided a somewhat striking example of the
necessity for verifying statements made even by great naturalists.

[1] Naumann, *Vögel Mitteleuropas*, xi. 108.   [2] See "Classified Notes."

Naumann asserts that the black-tern can catch no flying insects.[1] It
has, however, been seen—recorded by at least five observers—
catching insects that were flying above the surface of the water, and
doing it with grace and ease.[2]   Mr. Jourdain, who has frequently
watched the bird on the Continent, tells me that "no one can watch a
flock darting through a big rise of *Ephemerœ* without admitting that
they can and do take insects on the wing.   I believe they take
dragon-flies in the same way, but in some districts, such as the
Danube delta, these insects are found in millions and their capture
is very easy; consequently the birds would be more often seen
feeding upon them than in those parts of Germany where Naumann
lived."   Another record of their skill in insect-catching is given by
Heuglin, quoted in Hennicke's revised edition of Naumann's work,
the one quoted throughout the present book.

Besides hawking for insects in the air, the bird picks these and
other food off the surface of the water when in flight, dipping its
beak only.   It also alights for the same purpose on the water, and
has been once recorded as plunging in, like the Sea-Terns, with a
splash, completely submerging itself for an instant.[3]   It frequently,
when on the wing, picks insects from the water-plants, and has been
seen, like other sea-birds, following the plough for worms and grubs.[4]

Its usual call-note, which appears also to be an alarm-note, has
been syllabled by Naumann as *kliäh* (Anglicised *kleeèh*).   He
credits it with two other notes, *giek* and *kier*, the latter not unlike the
ordinary note of the Arctic-tern.[5]   According to Professor Patten,
"the note, which is frequently uttered, is shrill and powerful for the
size of the bird.   It sounds like *creek-crick*."[6]   H. Saunders renders it
as a shrill *crick, crick.*[7]   Neither of these two authorities alludes to the
second or third note mentioned by Naumann.

[1] *Vögel Mitteleuropas*, xi. 111 : "Können kein fliegendes Insekt fangen."
[2] T. A. Coward, *Fauna of Cheshire*, i. 419 ; Forrest, *Fauna of N. Wales*, p. 370; Stevenson, *Birds of Norfolk*, iii. 315 ; Yarrell, *History of Birds*, iii. 520; F. C. R. Jourdain *in litt.*
[3] A. W. Boyd quoted in Coward, *Fauna of Cheshire*, i. 419.
[4] Dresser, *Birds of Europe*, vol. viii.   [5] *Vögel Mitteleuropas*, xi. 110.
[6] *Aquatic Birds*, p. 374.   [7] *Manual of British Birds*, 2nd edit., p. 634.

Plate 98

Upper: Black-tern in summer plumage
Lower: Little-tern, showing erect attitude
of the male when feeding the hen

By A. W. Seaby

The enemies of the species are those of the Terns in general. It is attacked in the air by the faster Hawks, from which it seeks to escape by flying to an immense height, or by rapid twists and turns, in which its insect-catching renders it adept. Montagu, in his article on the black-tern, relates how "in a very hard gale of wind many terns were sporting over the water, when a peregrine-falcon passed like a shot, singled out his bird, and presently coming up with the chase, made a pounce, but the great dexterity of the tern avoided the deadly stroke, and took a new direction. The falcon, by his superior velocity, soon regained sufficient elevation to successively repeat his pounces, but at last relinquished the pursuit."[1]

The eggs are sometimes plundered by the usual egg-robbers, furred and feathered. Mr. Jourdain, again, tells me that the "Harriers, especially the marsh-harrier, will rob outlying nests, or even a small colony, but is at once mobbed by all the birds if it approaches a large colony. 'Furred' enemies must be very few: I fancy a dog would be driven off in many cases; only an otter would be dangerous, and that only in the few cases when the nests are on deep water."

COMMON AND ARCTIC-TERNS

[F. B. KIRKMAN]

In order to bring out more clearly what differences there are between the closely allied Arctic and common-terns, I propose to take these two together, apart from their three British congeners. Too little is known of the roseate-tern to give data for satisfactory comparison. Such information as we possess is summarised in a separate section. This leaves the little and sandwich-terns. As there is nothing much to be gained by emphasising the points of comparison between them, they are treated separately, attention, however, being drawn in each case to important particulars in which they differ from

[1] *Ornithological Dictionary.*

or resemble the Arctic and common. As the latter afford the standard of comparison, they are taken out of their order and placed first.

Though easy to distinguish when in the hand, chiefly by means of the marked difference in the breadth of the dark stripe on the inner web of the long quills (see the figure on p. 61), the Arctic and common-terns are so much alike in general appearance that they are sometimes confused when seen free in their native haunts. As incorrect identification has for result incorrect statements as to geographical distribution, it is worth while to note briefly the characters likely to aid the naturalist in distinguishing the species, when he is unable to handle them. The most reliable is the coloration of the bill. In the Arctic form this is blood-red, except for an occasional small black line on the ridge of the upper mandible at its tip, which may, for practical purposes, be ignored. The bill of the common-tern is dusky coloured for about a third of its length from the tip, the basal two-thirds being coral-red. The shorter tarsi of the Arctic constitute another difference, but it is of little use unless both species are under observation at the same time, except to those to whom they are familiar. Differences in the shade of the beautiful silvery grey of the plumage are of uncertain value for identification, because they vary for the eye with each change of light. By some the relative length of wing and tail is held to afford a reliable means of distinction. This is based on the assumption that when the common-tern is sitting or standing the tips of its wings reach either as far as, or beyond, the extreme tips of the tail, whereas in the case of the Arctic the wing-tips fall somewhat short of the tail-tips. There can be no doubt that, on the average, the common has, for its size, longer wings and shorter tail streamers, as the following figures show —:

	1. (H. SAUNDERS [1]).		2. (NAUMANN [2]).	
	Common.	Arctic.	Common.[2]	Arctic.
Length (full) . .	14·25	14·50 inches	30·6 to 35·30	34 to 39 c.m.
Tail	6·50	7·50 ,,	14·7 „ 15·8	17 „ 19 „
Wing	10·50	10·0 ,,	26·50 „ 27·10	26 „ 27 „

[1] *Manual of British Birds*, 2nd ed., pp. 648, 650.   [2] *Vögel Mitteleuropas*, xi. 128, 138.

Plate 99

Common-tern rising with fish

By A. W. Seaby

The variation shown by individual birds is, however, so great that this means of identification can only be regarded as of secondary or supplementary value. At the Farnes, where both species can be seen side by side, I have noted common-terns whose wing-tips fell distinctly short of the tail, and also an individual of each species side by side with the same relative length of wing and tail. The wing-tips of the Arctic may project beyond the tail. In one case quoted they did so by 1½ inch (38 mm.).[1] The extent of the variation in either species is well illustrated by Naumann's figures. According to him it has a sexual significance, the females having the shorter tails.

The common-tern's flight appears somewhat heavier than that of the Arctic; it seems almost as if, at each stroke, it were lifting weights attached to the tips of its pinions. This is, of course, chiefly noticeable when the bird is flying at a leisurely rate. It is noticeable to a greater or less extent in the case of all its congeners except the little-tern.

The species can further be distinguished by at least one marked difference in the notes they utter. The ordinary note of the common-tern is a long-drawn strongly dissyllabic *keee-yerrrr*; that of the Arctic a shorter *kerr*, occasionally *kerr-err*. Their remaining notes have yet to be closely compared. According to my observation they are much alike, but I have not heard the two species together, and base my opinion only on the fact that I have used much the same syllabic characters for their notes. From both I have heard a rapid *ptip, ptip ptip, ptip*. . . . What is the precise significance either of it or the *keee-yerrr* and *kerr* I am unable to say. They are possibly both used to express different degrees of alarm. I have heard the *kerr* and *ptip* uttered by Arctic-terns when taking their sudden periodic simultaneous flights from the ground after a recent disturbance. I have reason to think that the *ptip* is also a call-note, and may, in addition, have some sexual significance. There can be no doubt as to the meaning of the third note. This I find I have syllabled for the common-tern as

---

[1] *Vögel Mitteleuropas*, xi. p. 138. Or see *Journal für Ornithologie*, 1899, p. 884 (Schalow).

kikikerr or *kikiqwerr*. The *qwerr* (repeated) may be uttered alone, also the *kik, kik kik.* . . . For the Arctic, also, *kikikerr*, or *kerr, kerr, kikikikikiki kerr*, or *titwerr, titiwer, titiwer*. The variations were noted at different times, and are probably to be accounted for by variation in distance and direction. The birds used the note when fighting among themselves, attacking other species or an intruder.

It is when we compare two closely allied yet well-defined species, resembling each other so closely in external appearance and habit as do these terns, or, to take another example, the chiff-chaff and willow-wren, that the problem of the origin of specific variations assumes its most puzzling form. Seeing that the chicks of both Arctic and common-terns have black-tipped bills, one may assume that it is the Arctic which has diverged most from the common ancestor. If so, by what process or for what reason did the Arctic lose, except as a mere survival, the black tip to its bill, and why has it shorter legs and wings, and a narrower stripe on the primary quills? Why, on the other hand, has it become the larger species and, for its size, developed a longer tail?

The cause probably lies in the influence of differing environment, but it would be difficult to trace it either in the present habitat or present geographical distribution of the two species. The only important difference in habitat is that the Arctic shows a decided preference for salt as compared to fresh water, but it is only a preference, for the species has been found breeding on inland lakes in Ireland, Scandinavia, and Arctic America.[1] The common-tern is frequently found on fresh water even as far away from the sea as the Swiss lakes, and one of its specific names, *fluviatilis*, points to its liking for rivers. But that it shows a decided preference for fresh water is not clearly demonstrated. Some of its largest colonies are found by the sea, for instance at Romney Marsh in Kent, by the tidal estuary at Ravenglass, Cumberland, and at Muskeget Island off the New

---

[1] Ussher and Warren, *Birds of Ireland*, p. 323; H. Saunders, *Manual of British Birds*, 2nd ed., p. 650; A. L. V. Manniche, *Terrestrial Mammals and Birds of N.E. Greenland*.

England coast.[1] At the Farne Islands both species breed on the same stretch of sand and gravel. They also breed together at Walney Island in North Lancashire. The common-tern may once have been a purely fresh-water species, and the Arctic purely maritime, but to assume this does not help us to explain their divergent evolution.

When we turn to the difference in their geographical distribution we come upon one of the most remarkable recent discoveries in this branch of ornithology; it is that the range (summer and winter) of the Arctic-tern has a wider latitude than that of any other vertebrate animal. It has been met with not only among the ice-blocks on the road to the North Pole, at as high a latitude as 82°, but as near to the South Pole as 74° 1′ latitude, attracted there especially by the surface-swimming crustaceans that abound in the Antarctic seas during the southern summer.[2] It has been found in every ocean. Its range includes that of its congener, which lies well within the temperate and tropical zones. But, again, the fact that the Arctic, alone of the two species, ranges into the frigid zones, north and south, leaves us still in the dark as to the cause of its specific peculiarities. Yet unless the explanation is to be found in this fact, the search elsewhere seems hopeless, for between the present habits of the two species, to which we now turn, there is no difference that can be called important.

The common-tern arrives earlier at its summer quarters, the time varying with the season and the climate. It falls generally about the end of April.[3] The Arctic appears early in May. Both species migrate usually in bands, and, like the black-tern, in a leisurely manner, descending to fish in any promising water that presents itself. They journey both by day and night, and at a great height.

[1] *Auk*, xii. 33-48.

[2] *Ibis*, 1907, p. 345 (W. E. Clarke on the "Results of the Scottish National Antarctic Expedition"). The Antarctic birds may prove to be a local race or subspecies. See "Classified Notes," p. 63, footnote.

[3] For definite dates see the "Classified Notes." At Muskeget the earliest common-terns arrived during the years 1892-95 as follows:—1892, May 10; 1893, May 8; 1894, May 8; 1895, May 1 (*Auk*, xii. 33; xiii. 47). The late dates are due, no doubt, to the fact that the New England coast has a colder climate than the British Isles, though situated in a more southerly latitude.

On arriving at their breeding-grounds they drop, as it were, suddenly into it from the sky.[1]

The common-tern has been seen arriving at its summer quarters in pairs as well as flocks.[2] In respect to the Arctic I can find no definite information, but it is probable that both species, like many other birds, pair for life. The fighting that takes place may partly, if not wholly, be accounted for by the presence of males not previously paired or that have lost their mates.

Not having seen either species immediately after its arrival, I do not know whether there is a love-display different from that which may be seen later when the birds have settled down.   Probably there is not, for generally birds display in the same way before and after the actual winning of a mate, when such is necessary, which is not likely to be frequently the case with species that pair for life. The displays of the common and of the Arctic-tern, in the form that I have often witnessed it, are without any noticeable differences. They are most frequently associated with the feeding of the hen by the cock. The former, even before she has finally chosen the exact position of her nest, likes to stand near where it is to be and there await the return of her mate, who is expected to spend a large part of his time in zealous pursuit of dainties for the gratification of her palate. The presentation to her of the dainty, be it fish or crustacean, is accompanied by a ritual which, as far as the male is concerned, varies little. He alights, erects his tail stiffly, half spreads his wings so that they form with his back a more or less continuous surface, something like the top of a flat-iron, erects his neck just as stiffly as his tail, points his beak heavenward, and, with the fish hanging therefrom, patters round about his mate with precise little steps and a lofty air of doing something more than usually meritorious. The same attitude is assumed by the sandwich and lesser-terns, and is shown on Plates 102, 98. The behaviour of the hen varies according to her appetite. If she is in

[1] Naumann, *Vögel Mitteleuropas*, xi. 131, 141 ; *Auk*, xiii. 51 (G. H. Mackay).
[2] Naumann, *op. cit.*, p. 131; *Bulletin of the Nuttal Ornithological Club*, iv. 13 (W. Brewster).

no hurry she shows herself quite disposed to humour the cock's mood; she sidles up to him, her suppliant attitude in strong contrast to his, her neck contracted, her whole body in a lower plane, her beak turned up and sometimes opening and closing in anticipation. She takes the fish only when it is given her, and behaves, in fact, with marked decorum. If, on the other hand, she happens to be hungry, she hardly awaits for her mate to alight before she snatches the fish from his beak. This, however, in nowise prevents the cock, whatever his feelings, from duly performing his part of the ceremony.

I have never seen the presentation take place elsewhere than on the ground. Pairs may frequently be seen flying about in a wild chase about the sky, one holding the fish, but they suddenly descend to the ground for the actual presentation. At least such is my experience. Mr. A. W. Seaby once noted it take place on the water. The male hovered like a big silver butterfly over his mate, and she received the fish with upturned open beak. A pretty scene, except to the eyes of those for whom the part played in it by the fish is a matter of moment.

The display above described may be seen when there is no food brought. It is frequently seen when the pair is engaged in nest-building, also on occasions quite disconnected with any function. I have, for instance, seen two common-terns suddenly alight, both strike the erect posture, head up, tail up, flat-iron back, and then immediately fly off. Mr. Jourdain saw a pair of Arctic-terns display in the same way in the water. As in the case of other species, individuals may not infrequently be observed displaying when quite alone. I have, among others, a note of an Arctic-tern pattering about on a patch of sand by himself. Besides posturing in the usual way, he nodded his head in a private and confidential manner, as if engrossed by some matter of mighty import.[1] The general effect was marred by the extreme shortness of his little red legs. He had the

---

[1] My note reads "beak gracefully erected and lowered." This was written at the Farnes two years ago, at the time of watching. It conveys to me at present not much meaning. I am not, therefore, sure that "nod" is the right word to use.

appearance of waddling, and did not, therefore, do justice to the state of emotional exaltation in which he appeared to be.

On one occasion I saw two Arctic-terns facing each other, both in the attitude described, except that one had his head lowered so that the tip of the beak touched the ground. They seemed to be in an aggressive mood, and may have both been males. On another occasion two of the same species made an attack on each other. They both had the heads inclined, and bowed or nodded. It may be, therefore, that the display is used in a modified form as a sign of pugnacity.

Both species lay their eggs on the ground; the only exception to this rule that has come to my notice being provided by the common-terns that nest in the Danube delta[1] and in Holland[2] on floating masses of water-plants, and in Hungary on logs floating in rivers.[3] There is no satisfactory evidence to show that they differ in their choice of site. Naumann, who was very familiar with both, and was, with Nitzsch, the first to establish beyond doubt the specific distinction between them, held the view that the common-tern preferred gravel to sand, and avoided ground with any but the scantiest vegetation upon it.[3] This conclusion, based on his own repeated observations, made for the most part at least in Germany, is not borne out by evidence now available. It is possible that the species lays more often on gravel than sand, and it undoubtedly lays on bare rock,[4] but it certainly does not seek to avoid vegetation. On the vast shingle stretches of Dungeness, it is true, the number of nests on shingle greatly exceeds those placed elsewhere. Out of fifteen nests I examined twelve were on shingle, three on soil among herbage, one of the latter being among white campion at the foot of a foxglove. Among the sand-hills of Ravenglass, on the other hand, the great majority are placed

---

[1] R. B. Lodge, *Bird-hunting Through Wild Europe*, p. 268.
[2] Naumann, *Vögel Metteleuropas*, xi. 134, editorial quotation from Jourdain and Leverkühn.
[3] *Loc. cit.*
[4] Ussher and Warren, *Birds of Ireland*, p. 320; J. Harvie-Brown and H. A. Macpherson, *Fauna of North-West Highlands*, p. 312; Forrest, *Fauna of North Wales*, p. 373; C. J. King (Scillies), *in litt.* I have found nests on the bare rock at the Farnes.

season after season in the flats either on close-cropped turf or among the rough vegetation growing in the sand, chiefly marram grass. The minority are to be found on bare sand, and fewer still on shingle or gravel. On islets off the coast of Wales Mr. Jourdain found hundreds of nests among a luxuriant growth of sea-cabbage over 18 inches high.[1] Nests on close-cropped turf or among rough grass are to be found in Holland, in Ireland,[2] on Muskeget and Pennikese Islands off the United States' eastern seaboard,[3] in Mecklenburg,[4] also on the island of Borkum off the mouth of the Ems,[5] and, according to notes I made in 1905, on Walney Island in North Lancashire. (See photo, Pl. XLIII.) No doubt other places could be mentioned. On the Scillies the species often places its eggs in masses of seaweed, not infrequently within the tide-washed area, thus courting disaster.

The Arctic-tern, according to Naumann, prefers above all for its nest-site flat low ground covered with turf or other vegetation. This, he is careful to state, was the conclusion drawn only from his personal observation. Subject to this qualification, he thus makes a distinction between the nesting habits of the two species, for, as above noted, he thought that the common-tern usually nested on bare ground. This, we have seen, is hardly the case. Nor is it at all certain that additional information will support his view as to what is the usual nesting-ground of the Arctic. In Shetland its nests were often found by Saxby on the grassy flats.[6] In Jylland, Denmark, Mr. R. B. Lodge and Mr. Jourdain saw it nesting in scratchings in bare sandbanks in the fiords. At Walney Island, in 1905, of thirty nests examined I found twenty on sand (see Pl. XLIII.) or shingle. Two or three of these twenty were among the seaweed on the beach, and were washed away by a high tide. Of the remaining ten, nine were on sand among

---

[1] *In litt.*   [2] Ussher and Warren, *loc. cit.* Cf. *British Birds*, iii. 169, 201.
[3] *Auk*, xiii. 51; *ibid.*, xiv. 278.
[4] Wustnei and Clodius, *Vögel Mecklenburgs*, quoted by the editors of Naumann, *Vögel Mitteleuropas*, *loc. cit.*
[5] In 1869. See Droste-Hülshoff, *Vogelwelt der Nordseeinsel Borkum*, p. 230.
[6] Saxby, *Birds of Shetland*, 1874, 326.

thin newly sprouting marram grass, and one only among fairly thick
marram grass.   Here it was the common-terns that nested on the
vegetation-covered patches.   At the Farnes, in 1909, the great
majority were nesting on the beach on sand or shingle, only a few
in the adjoining sea-campion.   A few laid their eggs on the bare
rock.   At Rhos Colyn, in Anglesea, a large colony nest on rocky
stacks,[1] similarly in Shetland[2] and many localities on the west coast
of Scotland.[3]   Mr. H. C. Hart found nests in several places in the
Arctic regions on the bare shingle of the beach, and, in the same
region, they have been found in snow three inches deep.[4]

Enough has been said to show that much more evidence than we
have at present is required to establish beyond doubt that there is any
difference between the species in respect to the kind of nesting-site
each prefers.   In the evidence, such as it is, and viewed as a whole,
no marked difference is apparent.   It is worth adding that in estimat-
ing the value of evidence, it must be remembered that the ground
occupied by terns, in the latter part of the season especially, is not
necessarily that which they prefer, as persecution causes them to shift
their quarters.   They may leave what, from their point of view, are
desirable quarters for less desirable.

According to Naumann the common-tern always, and the Arctic-
tern usually, leave their nests unlined.[5]   Howard Saunders reverses
the statement, and makes the common-tern occasionally add a few
cross bents.[6]   Both authorities hold the unlined nest to be the rule.
This is, however, doubtful.   Unlined nests are no doubt frequently
found.   But so are lined.   Of about eighty nests of the common-tern
examined at the Farnes, Walney, Ravenglass, and Dungeness, I found
only two unlined.   An example of a well lined nest is given on
Pl. XLIII.   At the Wells colony in Norfolk, Mr. A. H. Patterson
found the nests lined " with few exceptions."[7]   Again, at Colchester

[1] *Zoologist*, 1905, 217 (Coward and Oldham).
[2] Evans and Buckley, *Fauna of Shetland*, 1899.
[3] F. C. R. Jourdain, *in litt.*
[4] *Zoologist*, 1880, 206 ; *Ibis*, 1877, 408.
[5] *Op. cit.*, xi. 134, 146.
[6] *Manual of British Birds*, 2nd ed., pp. 648, 650.
[7] *Zoologist*, 1905, 260.

Harbour, Mr. A. Hepburn[1] found few nests with no lining. Mr. Jourdain found nests of the common-tern in Holland all lined. Mr. Hepburn also made a discovery which probably explains much of the contradictory evidence on this subject in respect to both species. It was that the lining was accumulated gradually by the sitting bird. In the majority of cases a "slight" lining was recorded on the first finding of the nest, and a "thick lining" or a "big pad" on the visit a week later. This was particularly the case after a wet week, when a number of nests were made up with fresh material placed on top of the old. I have noted the same in the case of the gannet, and it is common, Mr. Farren tells me, to most or all the Waders, and no doubt also to all species with similar nesting-habits. The fact possibly explains why at the Farnes on May 29, when the Arctic-terns were just beginning to lay, I found very few lined nests among the thirty or forty I examined, and these few contained only some odd stalks that may not, indeed, have been put there by the birds, whereas of thirty nests of the same species examined at Walney at the end of June I found only two without lining. Some nests unquestionably remain unlined. The one shown on Pl. XLIII., from which young were hatched, is an example. The average proportion of lined and unlined has, however, yet to be established.

In the choice of material for the lining the species show no perceptible difference. They take as a rule what is near at hand, but why from the material available in one and the same colony certain individuals choose say pebbles, and others bents, or why, again, some dispense with a lining altogether, are questions that still beg a satisfactory answer.[2] Occasionally terns will neglect material near the nesting-site for less accessible. On the rock stacks at Rhos Colyn, Anglesea, where Arctic-terns breed, Messrs. Coward and Oldham noted that the only vegetation was lichen. Several made their nests

---

[1] *Zoologist*, 1910, 140.

[2] For a discussion of them see *British Birds*, i. 373 (W. P. Pycraft); ii. 78, 101 (F. B. Kirkman).

of this, but others used bents, and a few rabbit bones, both of which they must have carried from the adjoining land.[1] The most common material used by both species are bents, stalks, and a variety of vegetable matter, including occasionally twigs. Pebbles and shells are also used, and at Walney I found some nests of Arctic-terns made both of pebbles and bents, in one case there being a complete outer circle of bents. In the colony of common-terns at Wells, of which mention has been made, Mr. A. H. Patterson found nearly all the nests lined with "sixpenny sized pieces of cockle and oyster-shells." Rabbit bones appear to exercise an attraction on both species,[2] but they are only able to indulge the taste to a limited extent, owing to the natural disinclination of the rabbits to supply the material in large quantities.

On visiting a colony of Arctic or of common-terns, one does not, as a rule, find them all nesting near together, unless, as at the Farnes, the space is confined. Where the breeding-ground is extensive they may be found scattered in groups over a wide area, isolated nests or smaller groups being often met with in between the larger. At Walney or Ravenglass such groups breed in the flats among the sandhills, the latter forming natural boundaries between adjacent settlements. On the shingle stretches of Dungeness, where there are no such boundaries, the common-terns are found dotted in groups that may be a quarter of a mile apart or more. The two species do not mix, and are not, as a rule, found at the same breeding-place. A skerry has been known to be occupied by each species separately in consecutive years.[3] Among places where they can be seen together are Walney Island and the Farnes. At both the common are in a minority, or were so in 1905 (Walney) and 1909 (Farnes).

The building of the nest has two stages, preparing the saucer-

[1] Zoologist, 1905, 217.
[2] H. A. Macpherson, Fauna of Lakeland, p. 414; Zoologist, 1905, 217 (Coward and Oldham).
[3] F. C. R. Jourdain, in litt.

shaped hollow in the ground, and adding the lining. The latter, as we have seen, may be omitted. In order to make the hollow, the bird pushes its breast into the ground and turns slowly round, using its legs at the same time to scrape out the sand, gravel, or soil. I have often seen birds of both species engaged in this work. They are easily recognised even at a distance by their elevated white tails, which look as if they were planted in the ground. Four or five or more scrapes are made before the final selection is made. In the case of both species, again, the operation is often accompanied by the posturing already described in connection with the feeding of the hen by the cock. While one bird is turning round and round, the other may be seen standing head up, tail up, wings part spread, in an attitude of condescending attention. When the first bird steps out of the scrape, the other patters up to it, eyes it a second, and steps in and proceeds to rotate. The posture is assumed by both sexes, when, that is, they are interested. One will occasionally stand idly by while the other is busy. The working bird not infrequently pecks in a somewhat spasmodic fashion at bits of sand or other objects within reach, as it rotates, and throws them down by its flanks. On one occasion I saw a common-tern, presumably the male, descend with a fish, strike the usual attitude, then almost immediately step into a scrape and start rotating, the fish being still in his beak. His mate stood by posturing. After a repetition of the performance, both took flight.

The operation of lining the nest I have not seen, but judging from the usually careless appearance of the finished work, the material is simply brought in the beak and dropped into place, also sometimes out of place. It is no doubt subjected to the process of rotation above described.

The common-tern starts laying a little earlier than the Arctic, the exact date depending upon latitude and season. It generally falls towards the beginning of June, and in Scotland later. The former usually lays three, and the latter two eggs. In the case of both species, the sexes share the work of incubation according to Naumann. The down-clad chicks are hatched at the end of three weeks. They are dimorphic (see "Classified Notes"), the difference, in the case of the Arctic-tern, extending to the coloration of the upper parts and legs, and in the case of the common-tern to beak also. The difference between the two types of chicks of the latter species is said to extend even to their disposition, that with flesh-coloured legs being more lethargic than the one with red legs.[1] A curious fact, which I noted some years ago, is that the chicks of both these species, as well as those of the sandwich-tern and black-headed-gull, have a vestigial claw on the bastard wing. This will probably prove to be the case with all the *Laridæ*, if not the *Limicolæ* as well. Why it should have survived in the case of ground-nesting birds is difficult to explain. The only species in which wing-claws are habitually functional is the hoatzin of S. America, a tree-nesting bird, whose chick uses its claws, when alarmed, to clamber out of its nest away into the adjoining foliage. It loses the claws when fully fledged.

The coloration of the chicks, when they are on rough stony ground or shingle, is highly protective. The markings on the back and flanks still further aid concealment by serving to break up the outline of the body, so that it does not stand out against its background. The chicks of both species quit the nest almost as soon as they can run. Whether they ever subsequently return to it I am not certain, but think not. They leave it either to meet the parent coming with food or else to seek shelter from the sun, not from fear of human beings. During the first few days, indeed, the tern chick has not learnt fear, as the following incident, among many, will show.

---

[1] Common-tern: *Auk*, xiii., 1896, p. 51 (G. H. Mackay, whose observations were verified by R. Ridgway and W. Brewster). See also *British Birds*, iii. 169, Oct. 1909 (H. W. Robinson), for the same observations made independently. At Dungeness, in 1908, I noted that some chicks were a much darker buff than others. Arctic tern: *British Birds*, iii., 1909, 200 (N. F. Ticehurst). In July 1905 I photographed at Walney, in an Arctic-tern colony, two chicks, one grey in ground-colour, the other buff. All the adult terns I saw in the part of the colony in which these chicks were belonged to the Arctic form.

I found an Arctic chick one hot day lying on the sand some distance from its nest, and knelt down to take its photograph. No sooner did it see the shadow cast upon the ground by my body than it rose, and ran to nestle in the most confiding manner against the side of my leg. That the chick will readily quit the nest to find shelter is evident from an account sent me by Mr. C. J. King, who, as he lay hidden, saw a chick of the common species leave the nest for a shady spot. This it did time after time when put back. The peregrinations of the chick are sometimes determined by its parents. On one occasion I saw an Arctic-tern coax its chick to follow it by walking a little way from it, then back, then on. When it had thus led it near to the desired spot, it lay down, clucked much like a hen, and made with its body the snuggling movements of a bird settling down to brood. The chick ran quickly to it, and was soon under shelter. This occurred some distance from the nest.

On which day of its life the chick develops the instinct of fear I am uncertain, probably after three or four days. The change in its behaviour, when approached, is striking. It crouches until aware that it is detected, then runs away, using its short wings to help it over the inequalities of the ground. One young Arctic chick I cornered showed fight, uttering a little harsh note of anger, and pecking vigorously. The behaviour of individual chicks of course varies, some being more apathetic than others. The young of a common-tern, that might have been nearly two weeks old, allowed me to touch it before it got up and ran away over the shingle.

The young are fed chiefly on small fish by both parents. According to Wilson, quoted by Dresser and Sharpe, the parent common-terns "alight with the fish they have brought, and, tearing it in pieces, distribute it in such portions as their young are able to swallow."[1] It is more than doubtful that Wilson ever saw this take place, for not only is the beak of the tern ill-adapted for tearing flesh, but the chicks both of the Arctic and common species have no difficulty

---

[1] *Birds of N. America*, quoted in *Birds of Europe*, viii.

in digesting their fish whole. A chick may not infrequently be found with the tail of a fish longer than itself projecting from the end of its beak. The tail gradually disappears inside as the head portion becomes assimilated. The fish is either given directly to the young, or is dropped in front of it, sometimes from the air. Insects and worms are also brought, especially when stormy weather makes fishing difficult. On a wet evening, at Sule Skerry, Orkney, Arctic-terns have been seen in hundreds all over the island, hovering about six feet above the ground, every now and again darting down to the ground, seizing a worm, carrying it off to the chicks, scarcely alighting to put it into the wide-open bills, and then flying off for more.[1] When the young are on the wing, they still continue to be fed, receiving the food in the air, or else waiting for it on the shore.

The Tern's method of fishing is to hover, with rapidly vibrating wings and with beak usually down bent, some twenty feet over the surface of the water, where it looks much like a large white butterfly suspended on the end of invisible string; thence, when it has located its prey, it drops head-first, its wings half closed, and enters the water with a splash, emerging almost immediately with or without its fish. The extent to which the common and Arctic-terns submerge themselves depends no doubt upon the height from which they descend. Personally I have never seen either species disappear completely. But that the Arctic-terns do so occasionally was the opinion of Saxby, who wrote that in deep water "they would bury themselves completely out of sight."[2] Both species also take food off the surface after the manner of Gulls, that is, without alighting or wetting their plumage; they pause just above the water, pick up the object with the tip of the bill, and are off again at once.

Their usual method of picking worms and insects from the ground appears to be simply a modification of that adopted in fishing. They hover, alight for an instant, and are up again. They have been

[1] Annals of Scottish Natural History, 1904, 23 (J. Tomison).
[2] Birds of Shetland, p. 326.

seen following the plough.[1] They seem occasionally to hawk for insects in the air, like the black-tern. Saxby writes of Arctic-terns "skimming over the pastures on fine summer evenings" in pursuit of moths, and, strange to relate, common-terns are reported to have been seen hawking for flies with the swifts over Exeter Cathedral![2]

When their breeding-grounds are trespassed upon, Arctic and common-terns, like Gulls, swoop down upon the offender, utter a harsh scream, then rise to swoop again. Of the two species the Arctic-tern appears to be the more daring, at least where human beings are concerned; individual birds will occasionally, indeed, carry their resentment beyond mere protest. I have been violently struck on the head by one when handling its eggs. Another was seen by Mr. H. H. Slater to descend with such force as to break several of the eggs that a girl was carrying on her head in a basket. His companion, who was carrying eggs in his cap, was struck more than once on the bare head, blood being drawn.[3] I have never been attacked by common-terns when they had eggs, and, though often attacked—swooped upon, that is—have never been struck by them when with young.

The Tern's method of striking is, as far as I know, peculiar. It delivers its blow, not like most birds with the wings, or like the Skuas, for instance, with the feet, but *with the point of the bill.* I was made acutely aware of this fact when attacked by the Arctic-tern referred to above. Later I was able to verify the observation by examining young blackheaded-gulls that had been killed by both common and Arctic. I found, on removing the skin, that their skulls were perforated in different places as if a tin-tack had been hammered into them and then withdrawn. The extraordinary accuracy of the Tern's aim is not to be wondered at, seeing that their form of attack is, again, but a modification of their method of

---

[1] Naumann, *Vögel Mitteleuropas*, xii. 133, 144.
[2] *Field*, 1906 (July-Dec.), 250.
[3] H. H. Slater, *Birds of Iceland*, p. 110.

catching fish—the only difference being that the bill in the latter case serves to grasp instead of to pierce.

They attack all animals trespassing on their nesting-grounds. They will drive off dogs, horses, and cattle, also sheep, and I have seen them pursuing a hare over the shingle stretches of Dungeness. The large Gulls and the Crows they follow and harry mercilessly. According to Howard Saunders, a flock of Arctic-terns has been seen to mob and drown a hooded-crow. Their attitude towards the blackheaded-gull, their frequent neighbour, is one of watchful and suspicious toleration. The young of the same species they attack and kill in the way above described. This may be explained by the fact that the young gulls are markedly unlike their parents in coloration, the result being that their identity is mistaken. Moreover, the sudden emergence of a number of strange birds, to all appearance from the ground, in the middle of the breeding season is in itself enough to excite suspicion in the breast of any tern with a proper sense of responsibility. The Waders, such as the oystercatcher and the ringed-plover, which nest among the Terns, are left in peace. So also are their young, probably because they are born within the colony instead of invading it, like the young gulls, from outside as fledglings, and perhaps also because their plumage is not conspicuously unlike that of their parents.

The worst enemies that the Terns have are undoubtedly human beings, in particular those women misguided enough to imagine that their charms can be enhanced by wearing on their hats the faded and dismal remains of these birds, beautiful only when alive and free. For millinery purposes Terns are butchered each autumn by hundreds, as soon as they cease to be effectively protected. They are all the more easy to shoot because the flock, either moved by sympathy or curiosity, has a habit of hovering over its dead and wounded.

An enemy less to be feared is the Hawk tribe, from which the tern singled out for pursuit seeks to escape by rising to an immense height. The adult birds appear often to escape, the chief victims

Plate 100

Arctic-tern chased by Arctic-skua

By G. E. Lodge

being the young of the year.[1] Both species are harried by Skuas, which pursue them and compel them to drop the fish they are carrying. (Pl. 100.)

Their eggs, in spite of their vigilance, are occasionally robbed by Crows and Gulls, also by rats. In 1905, at Walney, hardly an egg of the Arctic-tern escaped the last-mentioned marauders, traces of whose feet and tails were to be seen leading from almost every nest to the nearest tuft of bent, where the broken shell told the story of the theft. The deep narrow furrow made in each case by the tail in the sand seemed to show that the rat had used this appendage as a support while it hopped away on its hind-legs, clasping the egg to its breast. That it escaped being murdered by the terns is remarkable. Where sheep and cattle have access to the breeding-ground, some eggs are, of course, trodden upon. Some again, as previously noted, are washed away by the tides from nests laid on or below high-water mark.

The unfledged young suffer, no doubt, chiefly from rain and cold. Those of the common-tern, born on river banks, are also liable to be drowned by floods.[2] From birds of prey they appear to be well protected by their protective coloration, when this assimilates with their nesting-ground, and by the watchfulness of the parent birds, which also, no doubt, serves to guard them from creatures that hunt by scent. But with regard to this information is lacking.

The Arctic and common-terns both begin to leave their breeding-places in August. Of the manner of their departure little is recorded by British observers, but Mr. G. H. Mackay has given some interesting details of the proceedings of the common-terns in 1895, before they left Muskeget Island, off the New England coast. From September 1st to 7th they appeared to leave the middle of the Island, and roosted on the outside beaches on the west and south. From the 15th to the 22nd they were seen collecting in large flocks, mounting upwards in a spiral course, then descending again. On the 26th and

[1] Cf. Naumann, *Vögel Mitteleuropas*, xi. 135, 146.　　　[2] Naumann, *op. cit.*, p. 146.

27th several very large flocks rose up in the air, until lost to sight. They were going in a south-west direction when last seen.[1]

From August to October those Terns which have left their breeding-places may be seen on many parts of our coast, and on migration inland. They quit us gradually for their southern winter quarters, some staying later than others, the former, presumably, being those that are migrating from the more northerly breeding-stations, British or other. Records are given in the "Classified Notes" of the arrival of some of our common-terns off the coasts of Spain in September. But many more such records are required in order to render possible a comparative study of the emigration and winter movements of those individuals of the two species that summer in the British Isles.

ROSEATE-TERN

[F. C. R. JOURDAIN]

Numerically this species is much the rarest of our British breeding Terns, and it is also far the most local. At the present time any systematic study of its habits is attended with the greatest possible difficulty, for it is only known to breed regularly at two localities in our islands. One of these is by no means easy of access, and probably contains not more than fifteen or twenty pairs, while the other and much larger colony is carefully protected. Moreover, as will be seen later, any attempt to watch this colony for any length of time would probably result in serious injury to the breeding prospects.

There are, however, various points around the coast where a few pairs occasionally breed. Till recently a few pairs almost always nested on the Farnes, but they were never common there, and have failed to hold their ground in spite of protection. What the exact cause of their disappearance is it is at present impossible to say.

[1] *Auk*, 1896, xiii. 51.

Apparently they are crowded out by their neighbours, the common-terns, but why they should not be able to hold their ground is not clear. Possibly the protection now extended to the Terns may be the indirect cause; for the common-tern is an earlier nester than the roseate, and on the rocky islets, where both species breed, the space available for nesting-sites is limited. As the colonies of common-terns respond readily to protection and tend to increase rapidly in numbers, it may be that the roseate-terns are gradually driven away by finding their breeding-sites occupied already. Against their natural enemies the habit of breeding in large colonies is a great protection. Even the larger gulls, bold robbers though they are, would hesitate to provoke the attacks of some hundreds of terns, intent on delivering their needle-like thrust with the closed bill from above. But the unfortunate wanderers, thus driven out, and prevented from nesting under the protection of their neighbours, if they breed at all, must found new colonies, where they are at the mercy of the first prowling gull which passes by.

The fecundity of this species is also less than that of its congeners. The normal clutch of the common-tern is three eggs, while two are much less usual. Among the Arctic-terns, it is true, clutches of two are more general and three are less usual, but with the roseate-tern the clutch ranges from one to two, while out of a large number of nests examined *in situ* by the writer, not a single one contained three eggs. It is of course quite possible that clutches of three occur occasionally: some instances have been recorded, but they must be regarded as quite exceptional. Mr. H. Noble, who has had considerable experience of this species, is inclined to consider one as being the normal clutch, sometimes two, and he also has never seen three (*British Birds*, iii. p. 90). My own experience leads me to regard two as the normal clutch in an ordinary season in the British Isles. In the colony of roseate-terns discovered by M. Blanc in Tunisia, the clutch consisted invariably of a single egg.

Another point which may possibly have some bearing on the

question is the preference shown by this species for recesses in the
rocks as breeding-places.   Heavy thunder showers will fill these
hollows with water, and I have not the slightest doubt that many eggs
are chilled and rendered infertile by this means.   In this case the
skin and egg collectors cannot justly be held responsible for the
diminution of the species : a few have no doubt been shot from
various colonies, but the thorough identification of eggs in a crowded
colony is a work of time and patience, and the number of authentic
eggs in collections from the British Isles is very small.

The roseate-tern arrives at its breeding-grounds after the common
and Arctic-terns, and does not appear till the very end of April.
Although nests may be found within a foot of those of one of the other
species, the roseate-terns tend to breed apart from their neighbours
and close to one another.   In a colony where all three species were
nesting, Mr. W. Bickerton noticed a preference for association with
the common rather than the Arctic-terns, while in the main colony the
island is divided between common and roseate-terns, and only a few
Arctic-terns breed on outlying rocks.   On the Farnes, however, Mr.
Kirkman noticed a pair in 1909 standing on the ground among the
Arctic-terns, and this was also the experience of Messrs. Cummings
and Oldham on the Skerries.   A very different state of things is
apparent, however, on visiting the great stronghold of this species in
the British Isles.   From afar off the bold rocky island stands out
black from the sea, but its top is white with a snowy cloud of terns,
and even half a mile off the din of their creaking notes comes to us
across the waters.   As the boat is steadily rowed towards the island,
the terns rise in hundreds and fill the air above us, until the rock is
deserted and every one of its tenants is on the wing.   It is by no
means easy to estimate their numbers under these circumstances, but
probably from three to four hundred birds at least are flying overhead, uttering an incessant chorus of creaking notes.   There is not
the least difficulty in distinguishing the roseate from the common as
they fly overhead.   The former is really a smaller bird, but, on the

other hand, its two streamers or outer tail feathers are longer and form an excellent distinguishing character. The flight of the two species is different, the wings of the roseate-tern being less elevated and depressed, and the flight in consequence appearing less laboured and more buoyant, while its wing-beats are slightly quicker. From time to time, as a bird passes within a few yards, the lovely rosy flush on the breast from which it derives its name is clearly perceptible, while the black bill (instead of red, as in the common-tern) gives an unfailing test when the bird is settled in front of one. The last is of course the best character when watching a bird on the nest, as sometimes the long tail feathers are hidden, but on the wing overhead the " streamers " and the characteristic notes at once attract attention. The notes are rather difficult to express on paper. Oswin Lee says that the call-note is a long-drawn " *krr-ēēē*," rather like that of the common and Arctic-terns, but much more shrill and prolonged, while, when disturbed at its breeding-haunts, besides the usual " *kĭk-kĭk-kĭk* " common to all the Terns, it has a long piping note, " *kēē-ēē-ēē*," almost like a whistle, readily distinguishable among the babel of cries raised by the accompanying throng of common and Arctic-terns. Messrs. Cummings and Oldham speak of the harsh " *craak* " of alarm and the call-note " *che-wick.* Mr. Bickerton describes the alarm-note as a harsh " *crrark-crrark*," while in my own notes it is described as harsher than that of the common-tern, and not dissyllabic with the fall on the second syllable, as in the well-known " *kree-aa* " of the common-tern, but prolonged on the same note. On landing we find that the apparently barren rock is covered on the top with a luxuriant growth of sea-cabbage and other plants, whose growth is no doubt stimulated by the droppings of the birds, which whiten the ground in all directions. The sides are, however, boldly scarped and worked into innumerable rocky hollows. Eggs of both species are lying about in scores: one must walk warily so as not to trample upon them, and even at the first hasty glance it is easy to see that there are two very different types among them. The first thing to be done, after a hasty walk round, is to conceal ourselves as far as possible in order that confidence may be restored and the birds return to their nests. As far as we can tell, the roseate-terns are chiefly confined to a part of the island where the rock is much broken up, and on the steep sides are many nests which appear to belong to this species. Not long after we have retreated to our hiding-place we can see the Terns beginning to settle on the far side of the island, but the clamour overhead goes on undiminished. As soon as one is on its eggs, its neighbours rapidly follow suit, and presently twenty or thirty birds are sitting in front of us. But without any warning the babel ceases, and in perfect silence the whole body rises in alarm ; presently the cries break out again, and gradually confidence is restored. This is a curious characteristic of all three species, which behave in exactly the same way : all rising with a simultaneous rush of wings in silence, and often apparently for no cause except sheer nervousness. At other times the cry of an oystercatcher will put every bird on the wing.

But now they are settling down again, and we can distinctly see the long streamers of a bird whose head is hidden in a tiny cave on the steep side of the rock. We mark the spot and turn our glasses to the next bird, and can clearly distinguish her black bill. A long and careful watch convinces us that no common-terns intrude in the territory of the roseates, but, on the other hand, there is no neutral ground between the two species. Within a foot or so of a sitting roseate in a cleft of rock sit two common-terns on the sandy top, so that practically one has to map out the limits of the two species. Curiously enough, on the Skerries this is not the case, and nests of the roseate are scattered about among those of the Arctic-terns, but the breeding habits of these two species differ less than those of the roseate and common-tern. After half an hour's work of this kind we can rise from our cramped position, of course disturbing every bird at once, and proceed to the ground we have been studying from a distance. One by one the marked nests are identified, and now we see that all those on which we have seen roseates sitting contain eggs of practically the same type. Curiously enough, although a smaller bird than the common-tern, the roseate lays a slightly larger and decidedly longer egg. The characters of the two have already been described in the "Classified Notes" and need not be repeated here, while the nesting-sites are usually very different, the common-terns frequently breeding in scratchings on the comparatively flat top of the island and among the luxuriant vegetation, while almost every roseate's egg is in some kind of crevice or recess or little cave on the steeper rock face. Most of the eggs are laid on the shingly floor of the recesses: here and there a few bits of dry and dead vegetable matter may be found casually arranged, but practically no nest is made. There are, of course, exceptions to these rules : thus Mr. Oswin Lee found a nest with the unusual number of three eggs on the Culbin Sands on May 20, 1887, which he ascribed to this species. The date is exceptionally early, and the number of eggs suggests a possible con-fusion with a common-tern's nest, especially as the latter were breeding there, and the eggs were laid in a depression in the sand.[1] Many of the large colonies of this species in other countries also are known to breed on sandy flats,[2] but here we are chiefly concerned with the habits of the birds in our own islands. Mr. Oswin Lee gives some interesting notes on the breeding-habits of these birds. He says that the male is very attentive to the hen while she is sitting, and often hovers in the air over her calling to her. He feeds her on the nest with small fish, and twice saw one carry a large sand-eel to the sitting bird, when they both devoured it, tearing it up and eating it in little pieces.[3] There seems to be no record of the male taking part in incubation, but naturally observations on this point are very scanty. Roughly we may estimate the incubation period at about twenty days, on the analogy of the two allied British species, but here, again, accurate information is lacking.

[1] Cf. also *Ootheca Wolleyana*, ii. p. 301.
[2] Cf. *The Auk*, xii. p. 32 ; Whitaker, *Birds of Tunisia*, ii. pp. 346-7.
[3] O. A. J. Lee, *Among British Birds in their Nesting-haunts*, i. p. 38.

A full and careful study of the plumage of the young from the down stage onward is given by Dr. L. Bureau in his "Monographie de la Sterne de Dougall," published in the *Proceedings of the Fourth International Ornithological Congress* (1905). The same writer also gives his observations on the breeding season and time of departure as observed by him off the coast of Brittany. Here the birds breed earlier than in Great Britain, and probably arrive earlier. The first eggs are to be found from about May 16th to 20th, whereas in Great Britain the more usual time for full clutches is from the first week in June onward, and many fresh eggs may be found, when the birds have not been disturbed, even in the second week of June. The first eggs were hatched in Brittany about the end of the first week in June, while both old and young, in cases where the colony has not been disturbed, leave the islands about July 15th. In cases where the first eggs have come to an untimely end, the stay may be prolonged even to the end of August, but by the first days of September all have disappeared. In one colony observed by M. Levesque they departed between July 25th and 28th.

After the breeding colonies have been abandoned, Dr. Bureau states that the birds do not forsake the neighbourhood altogether, but reassemble in large flocks, composed of old and young birds, on the outlying islets, and have been observed from the beginning of August up to the end of October. The latest date of which he has any record is that of two young birds obtained on October 22, 1896.

One last word on the habits of this bird. Mr. E. G. Potter (*Zoologist*, 1899, p. 83) suggests that the roseate-tern is accustomed to rob the other terns with which it breeds, when it is feeding young. The evidence quoted is chiefly second-hand, and in some cases suggests a confusion of species, but, if correct, gives a possible clue to the disappearance of small bodies of roseate-terns from large colonies of other species.

SANDWICH-TERN

[F. B. Kirkman]

The Sandwich-tern[1] is of stouter build, has a relatively longer bill and a shorter tail than any of the preceding species.[2] These three features, taken together, make it easy at once to recognise the bird in flight, even when seen at a distance. At close quarters, or seen through binoculars, it is, of course, identified by its black beak with yellow tip, and the striking velvety black mane-shaped crest, which gives it almost a martial appearance.

Its usual note, again, is distinctive—a sharp "*kirr-whit*," not to be mistaken when once heard. The alarm-note, uttered chiefly when intruders are near the nests, is not unlike that of the Arctic and common species. If heard at a distance it sounds to me like a somewhat disgruntled "*quk*," and, near to, like "*keek*," or "*kweek*," or "*qrreek*." It varies in pitch with the intensity of the bird's feelings. The species also utters the "*qwerr*" or "*qwarr*" noted in the case of the other species, but not, as far as my observation extends, in conjunction with the "*kweek*."

In some of its habits the Sandwich-tern differs markedly from its British congeners, the common, roseate, Arctic, and lesser. It arrives at its British breeding-haunts at the end of March or early in April.

[1] Named after Sandwich in Kent, where, in 1784, it was first observed.
[2] The relative length of the beak and body, as given by H. Saunders in his *Manual of British Birds*, 2nd ed., are—

	Body (with tail).	Beak.
Common-tern	14·25 inches	1·70 inches
Arctic ,,	14·50 ,,	1·60 ,,
Roseate ,,	15·50 ,,	1·90 ,,
Sandwich ,,	16 ,,	2·50 ,,

The relative length of body and tail are given by Naumann in vol. xi. pp. 129, 138, 154, 164, of the *Vögel Mitteleuropas* as follows :—

	Body (with tail to extreme tips).	Tail to extreme tips.
Common-tern	30·6 to 35·3 cm.	14·7 to 15·8 cm.
Arctic ,,	34 to 39 cm.	17 to 19 cm.
Roseate ,,	35·50 cm.	18·5 cm.
Sandwich ,,	36 to 37·50 cm.	15·30 cm.

thus from three weeks to a month earlier than they. It is sitting on its eggs when they are still choosing sites for their nests. Like them it arrives in flocks, which are said to be preceded by small advance parties, or even by single pairs.[1] Unlike them, it nests in "packs," the nests being often not more than a foot apart. The packs may vary in size from a few pairs to many thousands.[2] The largest in the British Isles are to be found on the Farnes, which is composed, however, of not more than a few hundred pairs. On the same breeding-ground more than one pack may be found. At the Farnes in 1909 there was one on the Inner Wideopens and two on the Knoxes. There are usually about half a dozen at Ravenglass, often some hundreds of yards apart. There is reason to believe that these packs or sub-colonies arrive independently of each other and at different dates, for at Ravenglass it was noted by Mr. W. Bickerton that the young in some were hatched out earlier than in others. The following are his records for four groups in 1906.[3]

Group.	Date.	No. of Nests.	Eggs.	Young.
I	June 1	11	14	
II	,, 1	62	60	36
III.	,, 4	14	22	
IV.	,, 4	17	14	16

At Ravenglass the species makes its nest in the marram grass on the sunny slopes of the sandhills, among the blackheaded-gulls, some of the latter actually nesting within the packs. Below, scattered over the flats between the sandhills, are the common-terns. On the Farnes the nests were in 1909 on ridges of sand, shingle, and sea-weed (Knoxes), and in the sea-campion well above the beach (Inner

[1] Naumann, *Vögel Mitteleuropas*, xi. 157. [2] *Ibid.*, p. 160.
[3] *Country Life*, Oct. 3, 1908, p. 445.

Wideopens). At both places the Sandwich nested at a higher level than their congeners. Near Ballina, in Ireland, a colony nested "on a low, flat mudbank scarcely above the level of the water."[1] They also lay on sandbanks and rocks.[2] In the nature of their nesting-sites they differ, therefore, little from others of their genus, except that, like the little-tern, they have not been found nesting on bare skerries. As far as is known they never breed inland.

It would be difficult to say whether their nests are more often lined or unlined. On May 19 to 24, 1909, I examined over a hundred at Ravenglass, and found them mostly more or less lined with marram grass. On the 30th of the same month I examined all the nests at the Farnes. In the large colony on the Knoxes, hardly a nest was lined. A few contained odds and ends, but there could be no certainty that these were placed there by the terns. The birds on the Inner Wideopens nested in the sea-campion, and either made a scrape in the loose bits or laid on them. The nests were thus more or less lined, but apparently by no effort on the part of the owners. In all these cases the nests contained eggs. The difference in habits between the two sets of colonies is probably to be explained by the nature of the sites. On the Knoxes the supply of nest material is very scanty; at Ravenglass it is abundant.[3] There is no evidence that the birds add material to their nests during the season, but it is not improbable that they do so, especially after rain.[4]

The eggs are laid in the early part of May, and vary considerably in colour (see "Classified Notes"). Both sexes share in the duty of incubation. I have seen them several times take each other's place on the nest. Both when doing this, and when the cock is feeding the hen, the curious displays, already described in the case of the

---

[1] Ussher and Warren, *Birds of Ireland*, p. 316.

[2] Naumann, *op. cit.*, p. 160: "trockene Sandwatten und vom Meere umgegebene Sandbänke, oder Felsen mit von Natur abgeplatteten Stellen."

[3] H. Saunders (*Manual*, 2nd ed., p. 644) records both lined and unlined nests. Dr. Patten (*Aquatic Birds*, p. 385) makes the unlined predominate. Naumann (*op. cit.*, p. 160) states that there are no lined nests, a statement in which his editors apparently concur.

[4] See p. 85.

common and Arctic-terns, may and often do take place. The favourite
attitude—wings half spread and flat-topped, beak turned upward, tail
sometimes also erect, air of high condescension—is shown on Plate 102,
drawn from life at the Farnes by Mr. Seaby. The supplicating attitude
of the hen is also shown, from which it must not be assumed that she
always adopts it. She sometimes snatches the fish from her mate's beak.
On the other hand, she has sometimes to exercise patience. I have
seen a cock suddenly fly off with the fish, just when the hen had risen
from the nest in the full expectation of a meal. She crept very humbly
back. It might have been a scene from the *Taming of the Shrew*.
On the left of the same picture two males are seen parading before
each other, or rather round each other, for what purpose I could not
divine, as the performance abruptly ceased, each bird going about
his business as if nothing had occurred. The drawing, as a whole,
gives a very lifelike image of the corner of a Sandwich-tern colony.
No birds are more charming to watch, the soft grey and white of the
plumage, the black velvety wind-tossed mane-like crests, the small
black legs pattering with precise little steps over the sand, the quaint
ceremonious bows—gentlemen of the old school these terns!—the
fluttering butterfly-like descents to the ground, the white grey wings
flashing in the blue of the sky, or over the deeper blue of the sea, all
combine to make a picture of which one does not tire, which one can
sit and watch by the hour, and which, among the dusty ways of men,
one can see again in mental vision, and rejoice in its freshness and
beauty.

The displays of the Sandwich before nesting operations commence
I have not had an opportunity of watching. According to an account
published in Ussher and Warren's *Birds of Ireland* (p. 318), the males,
at this period, strut about among the females, "their heads being
thrown back and their wings drooping, or almost trailing on the
ground. After a time, if there is no response from the females, which
generally look on at the performance with the greatest unconcern,
one of the males goes off for a time and returns with a sand-eel in his

Plate 102

Sandwich-terns, corner of a nesting colony

By A. W. Seaby

bill, after which he again struts about with wings and bill in the same position, offering the sand-eel from one to another of the females as he passes along unnoticed, until at last he meets one who accepts his offering, when he sits down to settle their arrangements for the season." If the interpretation of the facts here given be correct, it is clear that, in the opinion of the Sandwich-tern, the shortest way to the hearts of the fair is through that portion of their anatomy which is generally supposed among human beings to afford the promptest means of access to the affections of the male sex.

The chicks are hatched after about three weeks' incubation, according to records kept by the watcher at Ravenglass, who marked eggs in the nest, and by Mr. F. G. Paynter at the Farnes, who placed some in an incubator. The same period is given by Naumann.[1] The chicks vary considerably in coloration, especially with respect to the beak and legs. I have found chicks, just beginning to feather, with beaks either dusky hued, darker towards the end and yellow tipped or all yellow, or, again, orange-yellow, dusky towards the end and yellow tipped. A young bird, noted by Mr. Charles Oldham, which had acquired its first plumage, but was unable to fly, had the bill yellowish horn, and a chick three or four days old had the same dull vermilion with blackish tip.[2] Mr. Oldham or I have again found legs of chicks greenish black, purplish, liver-coloured, also lead-coloured, and legs of feathered young dark purplish and lead-coloured.

About the second or third day after hatching the chicks have completely disappeared from the nesting-area. On June 14, for instance, I noted at Ravenglass in one pack a certain number of chicks which were just hatched or hatching, also a few a day or so more advanced. The next day they all had gone. Once away from the nest they are not easy to find. At Ravenglass, in company with Mr. Oldham, I found a few in the long marram grass some distance from the nests. They were crouching in little self-made scrapes, and when disturbed ran away, making active use of their wings, or rather

[1] *Vögel Mitteleuropas*, xi. 65. [2] *Zoologist*, 1908, 168.

VOL. III.

arms, to help them over the inequalities of the ground. One, which tumbled into a small hole, put considerable pressure on its arms to help itself out. In this habit they differ from the chicks of the black-headed-gulls, which run with their wings closed unless hard pressed, when they open them, moved thereto by what seems an instinctive desire to take flight. The method of the little gulls is much better adapted to escape through marram grass than that of the little terns, whose outspread arms are a hindrance rather than a help. It is on shingle or rough stony soil that the latter find their arms of use, a fact noted already in the case of the Arctic-tern (p. 89). This points to the probability that such ground is their natural nesting habitat.

The chicks are fed by their parents on small fish, which are either put in their beaks or dropped in front of them. When they are fledged, they are fed in the air as well as on the ground.

The parent birds differ from the other British terns in their method of fishing, in that it is their usual habit to dive right under the water, and remain under an appreciable time—some two to three seconds. Like the others, they hover, before diving, over the water. This they do at various heights, from twenty to sixty feet more or less, their beaks being usually directed straight downward. When they have marked their fish, they drop suddenly in an oblique direction, using their wings, which remain partly open to direct their course. They appear to close them, or nearly so, immediately before entering the water, which they do with a splash. Like the other terns, they also pick food off the surface of the water, after the manner of gulls, that is, without alighting.[1]

The Sandwich-tern has not appeared to me as aggressive in defence of its eggs and young as the common and Arctic, though, like them, it will swoop down upon intruders. Naumann states that it attacks vigorously, and strikes them with the wings. If the Sandwich does strike with the wings, it differs in this respect from the common and Arctic, who, as already noted, strike with the point of the beak. I

[1] See p. 90.

have never been struck by a Sandwich-tern, though I have often handled their eggs and young, and I find this to be the experience of others. Its enemies are those of the rest of its genus.[1] Of the manner of its departure little is known. On quitting the breeding-grounds the birds do not necessarily move southward. A young bird of the year, ringed at Ravenglass when a nestling (June 30, 1909), was captured thirty to forty miles north of that place on August 20th.[2] It seems, therefore, that the Sandwich-terns linger about our coasts awhile before the southward movement begins, and this no doubt, as in the case of other terns, takes place in a leisurely manner.

LITTLE-TERN

[F. B. KIRKMAN]

The little-tern is peculiar in having the front part of the crown white in summer as well as in winter; its congeners have it in winter only. It is, further, easily distinguished from them by its smaller size, being not much larger than a song-thrush, except in length of wing. It makes up for its size by its aggressive noisiness, and may often be seen racing on the wing up and down the beach in a state of fussy alarm, ready to make a prompt attack on any crow or gull that happens to cross over its breeding-ground. What, no doubt, increases its general air of alacrity is the fact that its wings, being shorter than those of the larger species, are beaten more quickly, giving it, comparatively speaking, the appearance of being always in a hurry.

The usual notes uttered by the species when it sees its breeding-ground invaded, is a sharp excited *kweek*, uttered often in conjunction with a sound like *tik*, in varying combination, e.g. *kweek, tik, tik, kweek, tik, tik, tik,* or *kweek, kweek, kweek, tik, kweek, kweek*. . . . Both notes may be uttered separately. A familiar note, often heard as the bird flies

[1] See p. 92.   [2] *British Birds*, iii. 181.

speedily to the sea, is a rapid, equally excited *tiri-wiri-tiri-wiri-tiri-wiri!*
a sort of general announcement to the beach that the little-tern
is coming, and that he means to assert himself, if necessary.

The first arrivals reach our south coast in mid-April, but they
are not seen at the more northerly breeding-grounds till the end of
the month or early May.[1]

Though there are several colonies scattered up and down our
coasts, these are usually small, sometimes composed of only a few
pairs. The average number of pairs to a colony would fall far short
of that of the other British-breeding terns, excluding, of course, the
roseate. The same fact was long ago noted by Naumann.[2] The
marked preference for the open beach as a nesting-place shown by the
species, as compared with its congeners, may supply the explanation,
for not only are its chicks thereby much exposed to the effects of
bad weather, but its eggs are often washed away. At Dungeness I
have found chicks shivering on the beach on a sharp windy day in
July, and some of them did not survive the experience. Deaths from
exposure to rough weather are also recorded annually at Spurn Point.[3]
A good instance of the damage done by high tides to the eggs is
provided by the same colony. There the danger is so well recognised that, when necessary, the watcher is in the habit of removing
the eggs "a considerable distance inshore, and the birds easily find
them." [3] One may here remark, parenthetically, that, if it is possible
for birds to be astonished, certainly those at Spurn Point have reason
to feel so, for it is not the usual practice of human beings, when they
take eggs, to deposit them where they can be found by their rightful
owners. Another colony that nested on a shingle bank in the estuary
of the Dovey was less fortunate. Twice, in 1903 and 1907, its home
was flooded, and the eggs floated away on the water.[4]

Though it nests usually, in the British Isles at least, on the

[1] For the dates see "Classified Notes."
[2] There are exceptions. Seebohm records that near a ternery, Missolonghi (Greece), he
found the little much more numerous than the common, a discovery which he celebrated by
blowing two hundred and fifty of their eggs.
[3] *British Birds*, ii. 320 (Oxley Grabham).   [4] Forrest, *Fauna of N. Wales*, p. 376.

beach, the little-tern, like the common, nests also inland on the shingly banks of rivers. In Germany it is far more numerous in such places than the common.[1] It nests usually in shingle, or on sand among stones, broken shells, and bits of seaweed. Mr. W. Farren has informed me that in Spain he saw it nesting on the mud in the marismas. Naumann, who in his time examined hundreds of nests in Germany, never saw one on bare sand.[2] In the British Isles and in Holland,[3] however, nests on bare sand are undoubtedly found. (Pl. XLIV.) Occasionally, after a heavy gale, the eggs disappear, buried in the sand to a depth of several inches. At Spurn Point it has been noted that "the old birds almost invariably scratch them out again, and make all right."[4]

Naumann's view was that the little-terns avoided bare sand because their coloration would, by contrast with the yellow, render them conspicuous to their enemies. It is certainly true that a little-tern sitting in shingle or on sand among stones and broken shells is much less easily detected than when on bare sand. I have often, at fifty yards distance, watched, through a strong binocular, one after another fly on to its nest in shingle, and then, on putting away the binocular and looking long and steadily at the spot, have yet entirely failed to detect the bird, so well did its coloration blend with the vague outlines of the black, white, and grey stones.

The chief value of its coloration to the species lies, no doubt, in that it enables the birds to incubate without betraying the whereabouts of their eggs. If it was evolved by natural selection for this purpose, it is interesting to note that the natural selectors, four-footed or other, have now in the British Isles ceased to exert an important influence. The little-tern, though it nests usually where its coloration has protective value, is nevertheless among the least successful of its kind in the struggle for existence. It is much less successful, for example, than the common and Arctic, though they nest often on

---

[1] *Vögel Mitteleuropas*, xi. 124.   [2] *Op. cit.*, p. 125.   [3] F. C. R. Jourdain, *in litt.*
[4] *British Birds*, ii. 321 (O. Grabham).   See also Forrest, *Fauna of* N. *Wales*, p. 376, and Patten, *Aquatic Birds*, p. 398.

rocks, sand, turf, in campion, marram-grass, and other positions where their plumage has no protective value, but where they thrive as well as when they nest in shingle, which renders them as invisible as it does the smaller species.

The conditions that govern the struggle for existence in the case of the Terns, as of many other birds, have been altered by man. He is at once their chief enemy and their chief friend. He destroys their other enemies, and protects them from the more predaceous of his own kind in the breeding season, while he leaves them often to be slaughtered afterwards. Against him concealing coloration is no defence, not even as a protection to the eggs, for he has only to watch the bird alight on its nest, and carefully mark the spot to be sure of obtaining the clutch.

I know of one case, however, in which the little-tern profited by its conservative habits. This was at Walney Island in 1905. While, as noted on p. 93, the eggs of the Arctic-terns, laid inland among the sand-hills, were almost all devoured by rats, those of the little-terns down on the beach escaped. How far this was due to the concealing coloration of the latter species it would be difficult to say, for the rats might have in any case avoided the beach, even if the whole colony of Arctic-terns had moved thereto.

Naumann, whom I quote because he was very familiar with the genus under consideration, never saw a lined nest of the little-tern. The few stalks he occasionally found in them had the appearance of being there by accident and not design. Yet nests lined with shells (Pl. XLIV.) or small pebbles[1] are not infrequently found in some of the British breeding-places, and there is a record of one "lined with a few dry stems of grass.'[1] Sometimes the eggs are laid on a bed of shells, or bits of shells an inch or two deep. The shells are brought either from a distance or gathered from the immediate surroundings of the nest, in the latter case a distinct bare zone being perceptible between the nest and the encircling shell-bed.[2] Usually the eggs are

[1] Macpherson, Fauna of Lakeland, p. 418.   [2] Irish Naturalist, 1899, p. 191 (Patten).

laid in an unlined depression, and occasionally among stones without depression. It will be seen therefore that the little-tern, like its congeners, displays marked individual variations in its nest-building. It makes either no nest, or a bared depression in the ground, or a lined depression. Examples of all three classes may be found in the same colony. The nature of the lining also varies, but not so much as in the case of the other terns, for the reason, no doubt, that a smaller variety of suitable material is available on the beach than farther inland.

Like the common and Arctic, the little-tern usually makes several depressions or scrapes before it selects the final spot for its eggs. The depression is made by the simple process of pressing the breast down into the ground and rotating (see p. 87), the sand or soil being no doubt, as in the case of the above-mentioned terns, scratched out backward during the process with the feet.[1]

The eggs are laid usually in the latter part of May. The male, both at this stage and before, is assiduous in feeding the hen. When presenting his fish, he strikes the same remarkable postures as the terns previously described (Plate 98). The little-tern is, in fact, just as fascinating to watch as its larger congeners. By "watching" I mean not marching about over its breeding-ground, but sitting, with a good binocular in hand, some fifty yards away from the colony; sitting still, and for hours at a time, and day after day. It is only thus that one comes to realise how extraordinarily interesting the birds are, and how extraordinarily little we know about their intimate life. It is true that the same acts are performed over and over again, but seldom or never in exactly the same way. There is repetition with variation, and these variations are worth close and careful study.

The chicks are hatched in less than three weeks. Like those of the species of this genus already described, they vary considerably in

[1] The Sandwich also no doubt acts in exactly the same way, but I have never had an opportunity of seeing it at work.

coloration. Mr. Oxley Grabham has noted two distinct types at Spurn Point, one much yellower than the other.[1] I made a note of two chicks (Dungeness) which had the ground-colour of the upper parts sandy-*grey*, whereas some described (Thames estuary) by Mr. Hepburn had the same sandy-*yellow*.[2] Grey or yellow, they are equally well concealed by their coloration, when crouching among the shingle. It may be worth adding that the two chicks I noted had the throat sandy-grey, thus providing an exception to Naumann's statement that the throat of the little-tern is pure white in the down stage.[3]

The chicks readily quit the nest, impelled no doubt by a desire to seek shelter from sun, wind, or rain, under the shady or lee side of a stone or plant. The chicks and fledged young are fed on small fish in the same way as those of the larger species. The parents usually descend direct on to the spot where the chicks are, but I have seen them, when conscious of being under observation, alight some yards away, then fly up again to alight nearer the chicks, and do this three or four times before reaching their goal,—a very different proceeding from that of the ringed-plover, which is frequently found nesting near them, and which, after alighting in the neighbourhood of its nest, approaches it in a much more discreet fashion, by a series of irregular short runs and with an air of making for nowhere in particular.

The little-tern's method of fishing is similar to that of the common and Arctic. It hovers, as if suspended, several feet above the surface of the water, its tail spread and slightly deflected, its beak usually bent vertically downward; it then drops head first, the wings part closed, and enters the water with a splash, but without as a rule submerging its whole body. It seems indeed almost at once to assume the horizontal position and rise. When hovering, the birds have been observed to "work themselves backward several inches," so as to get in the right position for the plunge.[4] I have seen them, when

---

[1] *British Birds*, ii. 319.
[2] *Vögel Mitteleuropas*, xi. 120.
[3] *Zoologist*, 1904, p. 173.
[4] *Zoologist*, 1904, pp. 173-9 (Th. Hepburn).

dropping into shallow water, instead of diving, alight on the surface. They often make false descents, that is, they check themselves and take wing before reaching the water.

The little-tern appears, for the most part, to set out upon its southward journey to winter quarters in September. In this connection a curious fact is related as the result of an attempt to introduce the species on to the Farnes by placing eggs in the nests of the common and Arctic-terns. The little-terns were hatched out, and they departed southward in the autumn with their foster-parents, but they were not seen on the Farnes the following spring. Unfortunately these birds were not marked, so we do not know where they went.

[1] *British Birds*, ii. 821 (O. Grabham).

THE GULLS

[ORDER: *Charadriiformes*. FAMILY: *Laridœ*. SUBFAMILY: *Larinœ*]

PRELIMINARY CLASSIFIED NOTES

[F. C. R. JOURDAIN. F. B. KIRKMAN. A. L. THOMSON. W. P. PYCRAFT]

BLACKHEADED-GULL [*Larus ridibundus* Linnæus. Peewit-gull, sea-maw, sea-crow, patch, redlegged-gull, turnock; rittock (Orkneys). French, *goëland rieur*; German, *Lach-Möve*; Italian, *gabbiano comune*].

1. Description.—The adult blackheaded-gull may readily be distinguished at all seasons by the deep blood-red colour of the beak and legs, and the broad margin of white along the front edge of the wing. The sexes are alike, and there is a distinct seasonal change of coloration. Length 16 inches [406·40 mm.]. (Pl. 103.) The breeding dress differs from that of the winter only in that the head is covered—except for a rim of small white feathers behind, and partly encircling the eye—by a "hood" of dark "coffee" brown, which is assumed in the spring by a moult. The back and wing-coverts are of a delicate pearl-grey, the rest of the plumage pure white, but the breast, during the breeding season, may be suffused with a faint and evanescent rose-pink. The six outer primaries have a black terminal band and a white tip, the rest of the feather being grey, save the three outermost, which are white instead of grey. The white tips to these quills are soon lost or greatly reduced by abrasion. In the winter the dark hood is lost, but there is a patch of dark grey over the auriculars. The inside of the mouth is of the same hue as the beak and legs, but paler. In the young in down the ground-colour of the upper parts is of varying shades of buffish brown, with in many cases a strong rufous tinge. This is relieved by black markings representing disintegrated stripes. As with all disintegrating patterns, they vary individually, but a median stripe on the crown, and two longitudinal lines down the hind-neck and back can generally be traced, and commonly, also, a pair of lateral stripes. The down plumage is followed by what further research may show to be in part, at least, a mesoptyle plumage. In this the crown, nape, and sides of the head are greyish brown, the upper part of the hind-neck white, forming a half-collar, the free ends of which blend with the white of the fore-neck. The rest of the hind-neck, interscapulars and scapulars are dark brown—varying in intensity in different individuals—each feather having a margin of paler brown. The back is pearl-grey, passing into white on the rump. The median and minor coverts are dark brown, with paler brown margins: the marginals white: major coverts pearl-grey, washed with brown at the tips and with a tendency to a dark slate-coloured shaft-streak. Secondaries with a black patch in the distal half, leaving a greyish white margin along the lower part of the patch. Primaries with the terminal half black, and faint white tips to all but the three outermost; basal portion of the feather white. Primary major coverts white. Bastard quills white, with a black patch margined with white at the tips. Tail white, with a broad black terminal bar. The fore-neck and uppermost flank feathers buff coloured, rest of under parts white. Down feathers are adherent to the tips of most of the feathers of the upper parts at this stage, especially on the head, scapulars, rump, and tail. Beak pinkish grey or purplish. Later the plumage of the adult in winter is assumed, save that the minor coverts remain brown and the tail retains its black bar. In the following spring and summer the immature birds may still be distinguished by the same tokens. To what extent they assume the brown hood at this stage is still a matter for investigation. [W. P. P.]

2. **Distribution.**—During the breeding season this species congregates in colonies in marshy districts, often near the coast, but also occasionally far inland. In the south of England there are colonies in Dorset near Poole, in Hampshire, at Dungeness in Kent, near the Essex coast, while there is a large colony at Scoulton Mere in Norfolk and another at Twigmoor in Lincolnshire. Several colonies exist in Wales, the largest being near Lake Bala. Northward colonies exist in Yorkshire, Northumberland, Durham, Walney Island, at Winmarleigh Moss near Garstang in Lancashire, and in various localities in Cumberland, of which that at Ravenglass is the best known. In Scotland it becomes more general, breeding commonly on the lowland marshes and also in the hill tarns, but most numerously in the Clyde, Forth, Tay, Solway, and Moray areas, though found throughout the mainland and also in the Orkneys and Shetlands, but absent almost entirely from the islands off the West Coast. In Ireland it is a common resident, breeding in enormous colonies in some of the inland bogs, such as the Bog of the Ring, Tullamore, and absent only from the eastern and southern counties and the

marine islands.⋅ A small colony has, however, been recorded from one of the Blasket Isles. On the Continent its northern breeding-range extends to Southern Norway, and the south and east coast of Sweden, while it also breeds on the coast of Finland, Lake Onega, near Archangel, and in the Vologda and Perm governments, as well as in the Færoes. Southward it is distributed in suitable districts over the Continent to the Mediterranean, but there is no proof of its breeding in the Iberian Peninsula or South Italy, though it nests in Sardinia. In the Balkan Peninsula it breeds in the Danube valley but not in Greece. In Asia it is met with in the valleys of the Ob, Viliui, Lena, and Kolyma, as well as in Kamchatka, but apparently not in the Yenesei valley, and is said to breed in Transcaspia and Turkestan. Its winter range extends to the Mediterranean and North-west Africa, the Azores, Egypt, and the Nile valley to the Egyptian Sudan, the Red Sea, Persian Gulf, Northern India, China, Japan, and the Philippine Islands. [F. C. R. J.]

3. **Migration.**—The blackheaded - gull is found in the British Isles at all seasons, and is indeed abundant throughout the year all round our coasts, although withdrawing from many inland districts in winter. Nevertheless, the species is notably migratory; within our area a distinct movement towards the southern parts becomes obvious on the approach of winter, and some of the birds leave our shores for still more southerly quarters. To what extent a winter immigration from the north may exist is uncertain, but it is known that birds bred in eastern Europe visit the south of England on passage. Some light has already been thrown on the question by the recovery of "marked" birds. In the first place, we have the case of a young blackheaded-gull born and marked in Aberdeenshire in 1910, and killed near Bayon, Gironde, France, about January 18, 1911 (cf. Thomson, *Proc. Roy. Phys. Soc. Edin.*, vol. xviii. p. 217). On the other hand, we have a case of a bird not only remaining within our area but actually moving northwards within it : born and marked in Cumberland in 1910, found dead at Fraserburgh, Aberdeenshire, on February 20, 1911 (cf. Witherby, *British Birds*, vol. iv. p. 336). We may also summarise the results obtained by marking in North-eastern Germany, in that they partly affect our area. A number of blackheaded-gulls were marked as nestlings at a colony at Rossitten, at the south-eastern corner of the Baltic Sea. The subsequent recoveries of these are considered to indicate three routes. One follows the Baltic, North Sea, Channel, and Atlantic coasts to the Bay of Biscay ; another goes part of the same way, and then crosses by the Rhine and the Rhone to the Mediterranean, reaching the Balearic Isles ; the third crosses direct to the

PLATE XLIV

Photo by F B Kirkman

Unusually large nest of blackheaded-gull (Ravenglass)

Photo by F B Kirkman

Three nests of blackheaded-gull in marram grass (Walney)

Photo by Riley Fortune

Kittiwake's nest and nestling

Adriatic, reaching Southern Italy and Tunis (cf. Thienemann, *Journal für Ornithologie*, 1909, pp. 449-458, and plate viii.). Whether we accept this interpretation or not, it is of interest to note that the first of these supposed routes is represented in the British Isles by records from Great Yarmouth and the Isle of Wight respectively. At all times very gregarious, and probably mainly a diurnal traveller, although sometimes heard passing overhead at night. [A. L. T.]

4. **Nest and Eggs.**—As a rule the nest is placed on tussocks in marshy ground, but sometimes it is built on floating vegetation in water, and also at times on dry land in the neighbourhood of marshes. Large colonies are also found nesting on sandhills near the sea. Sometimes the nests are placed so close together that it is difficult to walk between them without treading on the eggs. At Twigmoor a few pairs build in trees, several feet from the ground (Seebohm, *History of British Birds*, iii. p. 311; O. Grabham, *Field*, June 7, 1902; R. Fortune, *Naturalist*, 1910, p. 95, etc.). The nest is a tolerably substantial one, built of dead sedges, reeds, grass, or other vegetable matter, which, according to Naumann, is contributed by both birds, but see p. 145. (Pl. XLIV.) The eggs are generally three in number, but four are occasionally found in one nest, and in some cases are almost certainly the produce of one hen. In large colonies extraordinary variation is found. Some thin-shelled eggs are bluish white in ground-colour, and are without markings or have all the colour concentrated into one huge blotch or zone of deep brown at the big end, but most eggs vary from stone colour to olive-brown in ground, and are blotched and spotted with purple-grey shell-markings and deep brown or blackish. In some cases the ground-colour ranges to bright blue, which, however, soon fades, and other eggs have a greenish ground. A rare type is distinctly erythristic, and has a warm reddish ground with brownish red markings. Zones, sometimes very distinctly marked, of light blue occasionally appear on these eggs (cf. *British Birds*, iv. pp. 317 and 220). (Pl. I.) Average size of 76 eggs, 2·2 × 1·46 in. [52·8 × 37·2 mm.]. The eggs are laid at intervals of a day, and are incubated for 20 days (H. S. Gladstone), 17-18 days (P. H. Bahr, *Home Life of Marsh Birds*, p. 54), or 22-24 days from laying of last (?) egg (W. Evans). Both sexes take part in incubation. The breeding season begins about the middle of April in the south of England, and about a week later in the north, while in Scotland the average time is towards the end of April. Normally only one brood is reared in the season, but when robbed the hen will lay again several times. [F. C. R. J.]

5. **Food.**—In a report made in 1907 by Mr. D. L. Thorpe and Mr. Hope to the County Council of Cumberland, it was shown, as a result of 100 post-mortem

examinations, that the staple diet of the blackheaded-gull consists of earthworms and insects. In only 14 was grain found, and fish in 9. As many as 42 % contained those agricultural pests, the wire-worm and crane-fly larva. One bird was found to have devoured 30 slugs. Mr. R. Newstead, in his report to the Board of Agriculture in 1908, found after examination of 80 pellets and four birds, that insects formed the great bulk of the food, and these, with only one exception, were either injurious to man or did not affect his interests. Mr. Newstead draws special attention to the destruction of crane-flies by this gull. On soaking one of their pellets in water "it was found to contain the remains of about 400 crane-flies and 1600 of their eggs; the latter had evidently been taken while yet in the body of the parent. Each pellet probably represented a single meal, and there can be but little doubt that each bird would make at least ten meals daily of these insects. If this were so, a single gull would be accountable for the enormous number of 4000 crane-flies and their eggs per day, making an aggregate of 28,000 per week. As the gulls flocked together in hundreds, the number of insects which they devoured may better be imagined than described " (*Food of some British Birds*, 1908, p. 85). The species also devours small crustaceans and molluscs. It appears only to take to grain and fish when its favourite food runs short, and the former only when left uncovered. No complaints appear to have been made by farmers who use the drill. It will occasionally eat the eggs of its own species, and possibly of other species. From the ascertained facts, it may be concluded that the blackheaded-gull is highly beneficial to the agriculturist, and that the harm it does to fisheries is small. The young are fed, doubtless by both parents, chiefly on insects and worms. One young gull, which I disturbed, left behind for my inspection sixty caterpillars and four beetles. Marine worms were disgorged at Walney by young gulls persecuted by terns. It is stated that they are fed also with sand-eels (*British Birds*, iii. 169). [F. B. K.]

LITTLE-GULL [*Larus minutus* Pallas. French, *mouette pygmée*; German, *Zwerg-Möve*; Italian, *gabbianello*].

i. **Description.** — The little-gull may at once be recognised by its small size, the French grey colour of the primaries, and the dark lead-blue colour of the under surface of the wing. There is a distinct seasonal change of livery, and the sexes are alike. Length 11 in. [279·40 mm.]. The adult, in nuptial dress, has the head and upper part of the neck black; the mantle and wings French grey;

the hindmost scapulars, secondaries, and primaries are broadly tipped with white. The rest of the plumage is pure white, suffused on the breast with a delicate salmon-pink. Beak reddish brown, legs vermilion. In winter the forehead is white, while the crown, nape, and hind-neck are French grey, like the mantle and wings, which are as in summer. Behind and below the auriculars is a small black patch. Immature birds have the upper parts brownish black—varying in intensity individually—variegated by bars of white, which form the tips of the feathers; these bars are broadest on the scapulars. The wing-coverts are dusky, with similar bars of white. The major coverts of the primaries are black, and the outermost primaries have their outer webs black, and a band of slate colour along the shaft of the inner web. The inner primaries are dark French grey, fading into white on the inner web, and with a subterminal transverse bar of black and a white tip. The under surface of the wing is white, not lead blue as in the adult. The tail, which is slightly forked, has a terminal bar of black. Birds of the second year are to be distinguished from adults by the broad black band across the wing, formed by the minor wing-coverts, and the black tips to the outer tail feathers. [w. p. p.]

2. **Distribution.**—To the British Isles this species is an irregular visitor in winter and on passage. Its breeding-range extends in the west to the islands of Rödby (1901) and Klaegbanken in Denmark, while in the Baltic it bred in N.E. Germany at the Drausensee (1899) and at Rossitten in 1902, and has recently re-colonised Gotland; while in 1901 it bred in S. Upland and probably also in Jemtland on the Swedish side. On the Russian side of the Baltic it nests in some numbers at several localities in the Baltic provinces, and at Karlö in the Gulf of Bothnia. Eastward there are large colonies on Lake Ladoga, a single colony near Archangel, and others in the Moscow, Kazan, Ufa, and Perm governments. In Asia it breeds from the Ob valley (64° N.) east to the Sea of Okhotsk. Its winter range in Europe extends to the Mediterranean and the northern coasts of Africa, from Marocco to Egypt, but in Southern Asia there is only a single record of its occurrence in India, in January 1859, and it apparently migrates westward to the Caspian and Black Seas. Stragglers have been recorded from the Færoes, New York State, Maine, Bermudas, and Mexico. [F. C. R. J.]

3. **Migration.**—A cold weather visitor from Eastern Europe, in irregular numbers. It occurs mainly on the east coast of Great Britain, less frequently on the south of England and the west of Scotland, and irregularly on the west of England and on the Irish coasts. September and October are the months in which it is most frequently recorded. Unusually large numbers visited the York-

shire coast in the winters of 1866 and 1868, and the whole east of England in 1869, and again, after heavy easterly gales, in February 1870. On the last occasion, an unusually large proportion of adults was recorded. In British waters the little-gull is occasionally seen in small parties, and is sometimes noticed associating with Arctic-terns (cf. Nelson, B. *of Yorks.*, 1907, p. 668). [A. L. T.]

4. **Nest and Eggs.**—Does not nest in the British Isles. [F. C. R. J.]

5. **Food.** — Chiefly aquatic and other insects and molluscs; also small fish (Naumann, H. Just, Koenig, Sandman). The stomachs of birds examined by W. Meves contained chiefly small fish, insects being found only in a few (Dresser and Sharpe, *Birds of Europe.*) [F. B. K.]

COMMON-GULL [*Larus canus* Linnæus. Sea or winter-mew, kitty, cob; white-maa (Orkneys), small or blue-maa (Shetlands). French, *goëland cendré*; German, *Sturm-Möve*; Italian, *gavina*].

1. **Description.**—The pearl-grey of the back, and the yellowish green of the beak and legs, at once distinguish the common-gull from its congeners. The sexes are alike, and there is a barely perceptible seasonal change of coloration. (Pl. 104.) Length 18 in. [457·20 mm.]. In summer dress the head, neck, tail, and under parts pure white, the back and wings pearl-grey save the primaries, which are black tipped with white save the two outermost feathers, which have a subterminal white spot. After the autumn moult the head and neck are more or less conspicuously streaked with ash-brown. In the juvenile plumage the feathers of the upper parts are of an ash-brown fringed with dull white, but the upper tail-coverts are white with ash-brown spots : the tail is white with a broad band of black across the terminal half, as in the young blackheaded-gull, from which, however, it may at all times be distinguished by its larger size and the totally different coloration of the rest of the plumage. The crown and under parts are mottled, and the flanks are heavily spotted with ash-brown. The young in down are variable in coloration, but have the head more or less conspicuously spotted and striped with dull black, and the back obscurely striped. Under parts white. [W. P. P.]

2. **Distribution.** — As a breeding species the common-gull is unknown in England and Wales, but nests commonly on inland lochs up to nearly 3000 feet on the mainland of Scotland, and also on many parts of the coast, but not in the north-east. On the west side its limits extend south to the Solway Firth, and it breeds commonly on the Inner and Outer Hebrides, Orkneys, and Shetlands. In Ireland

it is now known to breed in Co. Sligo (where it was first discovered in 1855), as well as in counties Mayo, Donegal, and Galway, and on one of the Blaskets (Co. Kerry). On the Continent it formerly nested on Texel, and a single pair bred in West Holland in 1909. Eastward it breeds on the East Frisian Islands (Rottum and the Memmert), the North Frisian Isles (Ellenbogen and Amrum), in some numbers in Jylland, Denmark; on the Langenwerder in Mecklenburg, and in the Russian Baltic provinces and the coasts of Finland. Throughout Scandinavia it is common, and in Russia its range extends from the Lapland coasts and the Kola and Kanin Peninsulas southward; while its most southerly breeding-places are at the mouth of the Don and in Transcaucasia and Transcaspia according to Buturlin. In Asia it breeds in the basins of all the great Siberian rivers up to about 67°-70°, and on the Amur to Kamchatka. In western North America it is replaced by an allied race. The winter range of this species includes the coasts of Central Europe and the Mediterranean basin, while it has occurred in the Canaries and in the Nile Valley as well as in the Persian Gulf, and has strayed to Iceland and Labrador. Asiatic birds range to Japan and China. [F. C. R. J.]

3. **Migration.**—Resident. To what extent migratory movement exists can only be conjectured, and all that can be said with certainty is that the species becomes generally distributed in winter, and visits England and other regions in which it does not breed. Immigration to Ireland from Great Britain is considered probable (cf. Ussher and Warren, B. *of Ireland*, 1900, p. 332). [A. L. T.]

4. **Nest and Eggs.**—This gull breeds by preference on low grassy islands, the grassy tops of low cliffs, and the sloping shores of lochs, avoiding precipitous cliffs and rocky coast-lines. The nests are also found high up in the hills, at an elevation of 2000 feet (Harvie-Brown, *Fauna of the Tay Basin*, p. 340a; *Fauna of the Moray Basin*, p. 215). They are generally to be found in colonies, and are substantially built of heather twigs, seaweed, dry bents, rushes, etc., as a rule, though occasionally very little material is used to line the nest-hollow. In Scandinavia and the N. Baltic it has been known to place its nest in a coniferous tree, and to take possession of an old hooded-crow's nest. (Pl. XLV.) The eggs are normally three in number, and the clutches of five and six eggs, which have been recorded from Sylt and elsewhere, are probably the produce of two hens. In shape they are a much broader oval than those of the blackheaded-gull. The ground-colour varies from ochreous to dark olive-brown, and the markings consist of spots, blotches, and streaks of deep blackish brown, with underlying ashy shell-marks. Some varieties have the ground pale greenish or even light blue, and in such cases the

markings are often almost obsolete. (Pl. I.) Average size of 85 eggs, 2·3 × 1·6 in. [58·5 × 41·6 mm.]. Nothing appears to be recorded as to the length of the incubation period in this species or the share of each parent, but probably it lasts for about three weeks. The breeding season in Scotland begins as a rule about the second week in May, sometimes a few days earlier, and though only one brood is reared, second and third layings may be found fresh throughout May and June where the birds are much harried. In more northerly situations, such as the Gulf of Bothnia and northern Scandinavia, the eggs are often not laid till June. [F. C. R. J.]

5. **Food.** — The species is omnivorous. Worms, insects, crustaceans, molluscs, fish (alive and dead), garbage, occasionally grain, small birds, and voles, all form part of its dietary. The young are fed by both parents, probably on worms, insects chiefly, but exact information is lacking. [F. B. K.]

HERRING-GULL [*Larus argentatus argentatus* Pontoppidan ; *Larus argentatus* Gmelin. Cob, silver-back, ladrum-gull (Devon), white-maa (Shetlands). French, *goëland argenté* ; German, *Silber-Möve* ; Italian, *gabbiano reale nordica*].

1. **Description.**—The adult herring-gull resembles the lesser blackbacked-gull, but is at once distinguished by its larger size and the pearl-grey colour of the back and wing-coverts, the flesh-coloured legs, and the yellow rim round the eyelid. The sexes are alike, and there is a slight seasonal change of coloration. (Pl. 105.) Length 24 in. [609·60 mm.]. As in the lesser blackbacked-gull, the scapulars, secondaries, and primaries are tipped with white. The outermost primary differs from that of the lesser blackbacked-gull in that the white area of the terminal portion of the quill is much larger, and crossed by a narrow sub-apical bar of black, while the penultimate feather is tipped white, and has a large white patch or " mirror " some distance from the tip, and in the closed wing concealed by the tip of the fourth primary. The innermost primaries are pearl-grey ; while the black area gradually decreases in the outermost primaries from the first (outermost) to the sixth, which has only a black spot near the tip of the outer web. After the autumn moult, the crown and hind-neck are streaked with ash colour. The juvenile plumage differs from that of the lesser blackbacked-gull in the markedly paler hue of the ground-colour, and in that the white marginal areas are not of uniform width but irregular, giving a more mottled appearance to the plumage. On the long inner secondaries this marginal area of white forms a dentate pattern. The inner webs of the inner primaries are much paler than in the young of *L. fuscus*, wherein all

the primaries are of a dark brownish black. As they approach maturity, the two species can more easily be determined by the appearance of the mantle colouring, but at this stage the young of *L. argentatus* have the major coverts brown "marbled" with greyish white, while in *L. fuscus*, of approximately the same age, these coverts are conspicuously barred with greyish buff. The adult dress takes four or five years for its completion, hence the immature phases cannot be accurately described, owing to the gradual character of the transformation. The young in down have the ground-colour somewhat paler buff than the young of *L. fuscus*, and the dark spots and bars on the head and neck smaller and more sharply defined, spots dominating : the markings on the throat are less conspicuous, but there is a more or less distinct black bar along the fore-arm which is not present in the young of *L. fuscus*. In very young nestlings of *L. argentatus* the markings on the back are fairly well defined. [w. p. p.]

2. **Distribution.**—This species is more generally distributed along the coasts of the British Isles than any other of our breeding gulls, and is found almost everywhere except where the coast is flat, as in Lincolnshire and East Anglia. From Kent to Dorset it is the only large species of gull which nests regularly, though a single instance of the breeding of the lesser blackback in Kent has recently been recorded. On other parts of our coasts the two species are to be found breeding side by side, not only on the mainland, but on all the principal islands, including the Orkneys, Shetlands, Færoes, and Outer Hebrides. In Ireland it is very common all round the coast, and one colony exists on a bog in Antrim. On the Continent it is said to breed on the Berlengas off the coast of Portugal, and also near Vigo, while it also nests commonly on the coasts of north-west France and the Channel Isles. Colonies are also to be found in the dunes of Holland and on the west and North Frisian Isles ; on Sylt especially enormous numbers used to breed. It is also not uncommon on the Jylland coast and in the Kattegat, and is abundant on the Norwegian coast to East Finmark. In the Baltic it nests on the Swedish coasts and by the great lakes ; also on the coast of Finland, Lake Ladoga, the Russian Baltic provinces, and on the northern shores of Germany. In the Mediterranean, and from the Black Sea and N. Russia eastward, and the Atlantic Isles, it is replaced by the yellow-legged form, *L. argentatus cachinnans*, but it breeds in Greenland, Baffin's Bay and Parry Islands, though replaced in other parts of N. America by *L. argentatus smithsonianus*. In winter its range extends to the Mediterranean basin and the coasts of N. Africa, but many stay with us. American birds winter in the Gulf of Mexico and the West Indies. [F. C. R. J.]

3. Migration.—In winter a considerable amount of wandering takes place, having a distinct southward tendency, although apparently quite irregular in character. Our exact knowledge of these movements is practically confined to what has already been ascertained by the new method of bird-marking. In the summer of 1909 many hundreds of young herring-gulls were "ringed" on Memmert (at the south-eastern corner of the North Sea), and seventy-one of these were recovered during the ensuing winter, mostly from the near neighbourhood of their birthplace, the farthest record being from a distance of about one hundred and twenty English miles (cf. Thienemann, *Journal für Ornithologie*, 1910, p. 632). But a different result was obtained from the marking of over a hundred young herring-gulls at two localities on the Aberdeenshire cliffs in the summer of 1910, as the following tabular summary of the record shows :—

Sept. 8, 1910	Saltfleet, Lincolnshire.
Sept. 13, 1910	Aberdeen Harbour.
Sept. 1910	Ryhope, Sunderland.
About beginning of Oct. 1910	Tayport, Fifeshire.
Oct. 3, 1910	Hunstanton, Norfolk.
About Oct. 12, 1910	Eden estuary, near St. Andrews.
Nov. 15, 1910	Aberdeen.
About Jan. 30, 1911	Near Manchester.

Two other records may also be quoted here. A herring-gull marked in the same season at Loch Aan Eilean, the Lewis, Outer Hebrides, was reported from Stornoway (Lewis) at the beginning of December 1910. And another, a bird in its first year's plumage, was caught at night near Aberdeen on October 3, 1910, and recaught on Burray, Orkney, on May 20, 1911 (cf. Thomson, *Proc. Roy. Phys. Soc. Edin.*, vol. xviii. p. 216; and *British Birds*, vol. v. p. 100). [A. L. T.]

4. Nest and Eggs.—This species is more adaptable in its habits than some of the gulls: in many districts it breeds in scattered colonies along the face of precipitous cliffs, but it is equally at home on the grassy tops of low islands in the West of Scotland, or on shingle-beds, while sometimes it may be met with breeding on the moors at some considerable distance inland, as in Northumberland, and in Holland I have seen the nests among low sandhills. Exceptionally the nest may be seen built on some deserted or ruinous building. The nest is rather bulky and neatly built, but there is considerable variation in this respect, some, presumably those of young birds, being very small and care-lessly constructed. The materials used are chiefly heather twigs, stalks of marine

PLATE XLV

Photo by W. Farren

Common-gull's eggs and nestling

Photo by F. Heatherly

Herring-gull's nest and nestlings

Photo by F. B. Kirkman

Nesting ground of lesser blackbacked gull (Farne Islands)

plants, dry seaweed, etc., lined usually with dry grass and an occasional feather, and both sexes take part in the work of construction. (Pl. XLV.) The eggs are normally two, or more frequently three, in number, but four have been occasionally found in a nest which had the appearance of a clutch, while in large colonies in-stances of five and even six eggs in a nest have been recorded by Leverkühn and others, but are almost certainly the produce of more than one bird. They vary a good deal in colour and markings : most eggs range from stone colour to olive-brown or greenish in ground-colour, but exceptionally they may be found with an almost white or distinctly blue ground. The markings consist of spots, blotches, or irregular streaks of dark umber-brown, and light inky purple shell-marks. In rare cases the markings are almost obsolete. A beautiful and scarce type, which, however, has not occurred in the British Isles, is the well-known erythristic or red egg, with sienna-red spots distributed over the surface on a warm whitish ground. (Pl. I.) Average size of 104 eggs, 2·7 × 1·9 in. [69 × 49·4 mm.]. Incubation is performed by both sexes, but chiefly by the hen, the cock-bird mounting guard close at hand. There is some discrepancy as to the period. An egg hatched under a hen on the 26th day (W. Evans), and Naumann gives the period as nearly four weeks, but Mr. Paynter estimates it only as 21 days from observations at the Farnes. The breeding season begins in the last days of April and the first week of May in Ireland and S. England, and rather later in Scotland; and though fresh eggs may be found as late as June and even July, it is probable that only one brood is normally reared. [F. C. R. J.]

5. Food.—Practically omnivorous. Fish (live and dead), small mammals, small birds, the young and eggs of larger birds, insects, worms, crustaceans, molluscs, garbage (floating and other), all form part of its dietary. The young are fed doubt-less by both parents, but exact information as to the nature of the food given is lacking. [F. B. K.]

LESSER BLACKBACKED-GULL [*Larus fuscus* Linnæus.
Cob, parson-mew; saithe-fowl (Shetlands). French, *goéland à pieds jaunes*; German, *Herings-Möve*; Italian, *zafferano*].

1. Description.—This gull, when adult, is readily distinguished from the great blackbacked-gull on the one hand and the herring-gull·on the other by its smaller size and by the slate-grey back and yellow legs and toes. The sexes are alike, and there is a slight seasonal change of coloration. (Pl. 105.) Length 22 in.

[558·80 mm.]. The head, neck, lower back and tail, and the whole of the under parts are of a pure white, while the back and wings are black, relieved by white tips to the scapulars, secondaries, and primaries. The white tip to the outermost primary is extremely small, but below this is a large subterminal white spot or "mirror." By the end of the summer the tips of most if not all the primaries have been lost by abrasion. The beak is yellow, with a red spot at the gonys of the mandible ; the rim of the eyelid is vermilion, the iris pale yellow, and the legs and toes are yellow. After the autumn moult the white of the crown and hind-neck is marked by ash-brown striations. The juvenile plumage is of a sepia-brown, dull white margins to the feathers giving a mottled appearance to the upper parts, save the head and neck, which are obscurely striated, while the under parts are ash-brown. The beak is of a dark brown colour, the iris dark brown ; there is no orbital ring, and the legs and toes are brown. It is to be distinguished from the herring-gull by the conspicuously smaller beak, and the fact that the pale margins to the feathers are narrow and of uniform width, while the ground colour is much darker : the marginal pattern of dull white is especially well shown, for the purposes of comparison on the long inner secondaries. The adult dress is gradually assumed by a whitening of the under parts, rump, and tail, and the gradual darkening of the back. But the process of transformation takes about four years to complete. Hence immature birds cannot be accurately described, owing to the gradual transformation. The black, white-tipped primaries do not appear till the assumption of the adult dress. The young in down are buffish grey with irregular dusky bars, and spots on the sides of the head and crown, a long "horse-shoe" on the throat, and obscure, irregular bands on the back, representing disintegrated stripes. [w. p. p.]

2. **Distribution.**—In the British Isles this species is less widely distributed than the herring-gull, for not only is it absent from the flat shores of Lincolnshire and East Anglia as a breeding species, but, with the exception of a solitary nesting record from Kent, and a few pairs breeding among the herring-gulls in Hampshire, it is hardly known to nest along the shores of the English Channel east of Devon. In Wales there is a colony on a bog ten miles from the sea, and in Cumberland, and to some extent on the Northumbrian moors also, it is a common breeding species inland as well as on the coast. There is also a large colony on the Farnes, and it is plentiful up the west coast of Scotland and the Inner Hebrides, but less numerous on the Atlantic side of the Outer Hebrides and the east coast of the mainland. Northward it nests in the Shetlands and Orkneys, as well as in the Færoes. In Ireland it is much less common than the herring-gull, but has been found breeding

both on inland lakes and bogs as well as on marine islands. It breeds both on
the Scillies and the Channel Isles. On the Continent it breeds on the western
coasts of France, but it has not been found nesting on the coast of Portugal,
though Lilford and Irby state that a few breed on the rocky island of Alboran,
near Marocco, and a pale form is said to breed on the Azores. It is common
on the Norwegian coast, and also in Sweden and the shores of Finland, but
scarce in the Russian Baltic provinces. In Russia its breeding-range also
includes the Murman coast, the White Sea, Lakes Ladoga and Onega, and the
St. Petersburg government, but further east it is replaced by *L. fuscus affinis*. In
winter many stay with us, but others range southward to the Mediterranean,
where they visit the N. African coast from Egypt to Marocco, and also occur
on the west side of Africa to the Canaries, Senegal and Bonny. It also visits the
Red Sea and Persian Gulf. [F. C. R. J.]

3. **Migration.**—A resident and a summer visitor. In autumn a regular
southerly movement takes place, and the northern coasts of our islands are practically forsaken. That the movement extends beyond the southern limits of our
area is shown by the case of a young bird of this species, marked on the Farne
Islands, Northumberland, in June 1909, being recovered in the early winter of the
same year near Olhao, Portugal (*Country Life*, November 27, 1909, and December
11, 1909). On our southern coasts the species is found all the year round, although
more generally distributed in winter. [A. L. T.]

4. **Nest and Eggs.**—Although occasionally nesting on cliffs and broken
ground, this species shows a decided preference for the vegetation-covered tops of
islands and moors or bogs inland. The nest is much like that of the herring-gull, but
it is not so neatly finished, and is made of similar materials, heather twigs, moss, bits
of seaweed, and dead grasses. (Pl. XLV.) Naumann implies that both sexes take part
in providing material. The eggs are normally two, or more usually three, in number,
and clutches of four and five are probably due to two hens laying in one nest.
They vary a good deal in colour, but usually range from light stone colour to deep
olive or umber-brown, with blackish brown spots and blotches and underlying shell-
marks of light inky purple. Greenish tinted eggs are rarely found, and are not
nearly so common as with the herring-gull, but a pale blue variety, with almost
obsolete markings, is sometimes met with. (Pl. I.) Average size of 100 eggs,
2.63×1.84 in. [66.8×46.7 mm.]. They are thus on the average smaller than the
eggs of the herring-gull, but the measurements of the two species overlap. Both
sexes take part in incubation, which, according to Mr. Paynter, lasts for 21 days,

but confirmation is desirable. The breeding season begins about the first week in May, and, where the nests are much harried, fresh eggs may be found through May and June, but apparently only one brood is normally reared. [F. C. R. J.]

5. **Food.**—Omnivorous. Food practically the same as that of the herring-gull. [F. B. K.]

GREAT BLACKBACKED-GULL [*Larus marinus* Linnæus. Cob, parson-mew, saddle-back; baagie (Shetlands). French, *goëland à manteau noir*; German, *Mantel-Möve*; Italian, *mignaiaccio*].

1. **Description.**—This gull can be recognised at once from its congeners by its much greater size, and the dark, slaty black of the back and wing-coverts. Like the lesser blackbacked-gull, it has a vermilion rim to the eyelid, and, like the herring-gull, it has flesh-coloured legs and toes. The iris is yellow. The wing, during flight, shows a white border along both anterior and posterior margins. The sexes are alike, and there is a slight seasonal change of coloration. (Pl. 106.) Length 30 in. [762·00 mm.]. The scapulars and secondaries have conspicuous white tips: the two outer primaries are peculiar in that the white tips extend back over a much larger area than in either the lesser blackbacked or herring-gulls, while the three succeeding feathers have a large white area at the tip crossed by a subterminal black bar; the rest of the primaries have white tips. The head, neck, rump and tail, and under parts are pure white. The juvenile plumage resembles that of the lesser blackbacked-gull, from which it can always be distinguished by its much larger size. The adult plumage takes about four years to develop, and follows a course exactly similar to that of the lesser blackbacked and the herring-gull. The black mantle begins to make its appearance at the end of the second year; later, the white tips begin to appear in the primaries, and the black terminal portion of the tail becomes "invaded" by white. Probably the fully adult dress is not attained till the end of the fifth year, but precise information on this point is still wanting. The downy young are ash-grey above, mottled with blackish brown spots and bars, which are most conspicuous on the head and neck; the under parts are white. [W. P. P.]

2. **Distribution.**—As a breeding species this fine bird is entirely absent from the east of England, while only a few pairs breed in Dorset, Devon, the Scillies and Lundy Island. In Wales, also, many of its former breeding-places on islets in lakes are now deserted, but some still nest on the mainland and in Anglesey, and this is also the case in the Lake district in one locality. In Scotland, however,

PLATE XLVI

Photo by C. J. King

Great blackbacked-gulls and nest (Scilly Isles)

Photo by C. J. King

Castings of the great blackbacked-gull : two shearwater's heads (upper) : puffin's head (lower) and two balls of feathers (Scilly Isles). (See p. 167)

it becomes more numerous, and not only do single pairs nest on the islands and cliff-tops, but in some cases colonies of twenty or more pairs may be found breeding together on the top of some lofty stack, especially in the Outer Hebrides, where, as well as in St. Kilda, the Orkneys and Shetlands, it is locally common. It is resident in small numbers on the Irish coasts, breeding on the tops of the stacks, especially in the west, and there is one very large colony of at least 100 pairs off the coast of Mayo. A few nests have also recently been recorded on islands in lakes inland (*British Birds*, v. p. 139, etc.). Outside the British Isles it breeds in the Færoes and Iceland, while on the Continent a few nest in the north-west of France, but there is no other known breeding-place nearer than the coasts of Norway and Sweden, and on the inland lakes of Sweden, as well as on Gotland, the Russian Baltic provinces and the south-west of Finland. In North Russia it nests on the Kola Peninsula, the White Sea, the Kanin Peninsula, Kolguev, and the Petchora delta. In N. America it breeds in Greenland to lat. 68°, and also in Labrador and the New England coast. The winter range extends to the Canaries and Azores, as well as to the Mediterranean, Black and Caspian Seas, while on the American side it visits Bermuda and Florida. [F. C. R. J.]

3. **Migration.**—Resident. In winter the species becomes more generally distributed, and there is some evidence of a regular migration with a southerly trend. Nothing more definite is known. [A. L. T.]

4. **Nest and Eggs.**—The site of the nest is generally chosen with a view to safety, and when found singly is often placed on some small islet in a loch, or on the top of some isolated rock. Other species of gull are kept at a respectful distance, but it does not object to the society of its own kind, and on the tops of some of the more inaccessible stacks in Scotland and Ireland large numbers may be found breeding close to one another. Thus on the Lyra Skerry in the Shetlands about twenty-seven pairs breed together, but the numbers in this colony are far eclipsed by those in the Irish colony referred to above. The nest resembles that of the herring-gull and lesser blackback in appearance, being built of heather twigs, seaweed, but more especially grasses torn up by the roots, thrift, etc. (Pl. XLVI.) The eggs vary from two to three in number,[1] and normally range from pale stone colour to brownish in ground-colour, boldly blotched with umber-brown and different shades of ashy grey. Compared with eggs of the herring and lesser blackbacked-gulls, they show much more of the ground-colour on an average, and are more boldly marked, while a variety with bluish ground and very sparingly

---

[1] H. J. Pearson once found five eggs in a nest in Norway.

marked, or with only one or two bold blotches, is not uncommon. The erythristic type described under the head of herring-gull also occurs in this species, but not in the British Isles, and Dr. Ottosson records a melanistic variety of the egg. (Pl. J.) Average size of 50 eggs, 3·04 × 2·13 in. [77·4 × 54·1 mm.]. Both sexes take part in incubation, and the period is given by Faber as 28 days. The breeding season in the British Isles seems to be rather irregular : as a rule the first eggs are found early in May, but Mr. J. J. Dalgleish records a clutch on April 19 near Ardnamurchan, and many pairs do not lay apparently till the second or third week in May. Only one brood is reared in the season, but second layings may be found in June. [F. C. R. J.]

5. **Food.**—The species is omnivorous. Its food is practically the same as that of the herring-gull, except that its greater strength enables it to prey upon larger mammals and birds. (See for details p. 167; also Plate XLVI.) According to Saunders, it will attack sickly ewes and weak lambs, and it makes short work of wounded ducks and game-birds. The young are fed by both parents, but details of the nature of the food given are lacking. From a young bird in down an eel 18 inches long was extracted ! (*British Birds*, v. p. 88.) [F. B. K.]

GLAUCOUS-GULL [*Larus glaucus* Brünnich; *Larus glaucus* Fabricius. Burgomaster, Iceland scorie (Shetlands); Great white-winged gull (Devon). French, *goëland bourguemestre* ; German, *Eis-Möve*; Italian, *gabbiano glauco*].

1. **Description.**—This species recalls the herring-gull, but may at once be distinguished therefrom, when adult, by the entire absence of black on the primaries. The sexes are alike, and there is no marked seasonal change of coloration. (Pl. 107.) Length 29 in. [736·60 mm.]. The mantle and wings are of a delicate pearl-grey ; the primaries pale grey fading into white at the terminal third ; the rest of the plumage is pure white. The beak is yellow, with a patch of orange on the angle of the lower jaw ; the rim of the eyelid is vermilion, the iris yellow, and the legs and toes are bright pink. The juvenile plumage is of a pale buffish ash, but the mantle and wings are of a rich cream colour, relieved by loops of dark greyish buff. The greater wing-coverts and quills are of a buffish ash ; the tail cream colour, with darker irregular transverse bars. Immature birds have the head, neck, and interscapulars dull white, with pale brownish ash striations. The scapulars and wing-coverts cream colour, with broad submarginal loops of pale brownish ash. The secondaries are tipped with white, and mottled subterminally with pale brown, the rest of

the feathers being faintly mottled with brownish ash. The primaries are of a dull creamy white, with a greyish tinge, the innermost tipped with white, and with subterminal arrowheaded markings of brownish ash. Tail dull white, heavily marbled with pale brownish ash. As the bird nears maturity the markings disappear and the cream colour fades to white, so that in fourth year birds the whole plumage is for a short time pure white. Birds in this stage were regarded as representing a distinct species,—*Larus hutchinsi*. The young in down has the head dull white, with dusky spots on the crown, and lines and spots of the same hue on the sides of the head. The rest of the upper parts are pale ashy brown, with obscure lines of darker brown along the back and on the elbow ; under parts white. [w. p. p.]

2. **Distribution.**—This is a circumpolar species, breeding in the high north of both Old and New Worlds. Its nearest breeding-places to our shores are Iceland, Jan Mayen, Spitzbergen, Franz-Josef Land, the Murman coast, Kanin Peninsula, Kolguev, Petchora delta, Waigatz, and Novaya Zemlya. In Asia it breeds along the north of Siberia, and on the New Siberia Islands eastward to Tchukchi-land. In the Western Hemisphere it breeds commonly in Greenland, and sparingly along the coast of Labrador south to Newfoundland, as well as on the shores of Hudson's Bay and the Nearctic Archipelago, but in Bering Sea and Alaska it is replaced by the Point Barrow or western glaucous-gull, *L. barrovianus* Ridgway. Its winter range normally extends to the northern part of our islands, but exceptionally it ranges south to the Straits of Gibraltar, the Mediterranean and Black Seas in Europe, and the Caspian and Japanese Seas in Asia, while in N. America it has been recorded in winter from Florida, Texas, and the Mississippi valley, and on the Pacific side from California. [f. c. r. j.]

3. **Migration.**—A winter visitor. It is of regular occurrence on our northern shores, but always rare in Ireland and on the south and west of England. In severe winters it is more abundant; a notable season was that of 1871-72 (cf. Harvie-Brown, *Fauna of Tay Basin and Strathmore*, p. 339). Occasional examples remain in Shetland till June, but do not breed there. Over fifty have been seen together in Orkney in December (cf. *Annals Scot. Nat. Hist.*, 1902, p. 197). [a. l. t.]

4. **Nest and Eggs.**—Does not breed in the British Islands. [f. c. r. j.]

5. **Food.**—The species, like its congeners, is omnivorous (see p. 125). [f. b. k.]

132 THE GULLS

ICELAND-GULL [*Larus leucopterus* Faber. French, *goëland leucoptère*; German, *Pôlar-Möve*].

1. Description.—The Iceland-gull is to be regarded as a miniature of the glaucous-gull, but has relatively longer wings. It is to be distinguished from it at all stages of growth by its greatly inferior size and the coloration of the orbital ring and legs and toes, which are flesh-coloured. (Pl. 105.) Length 22 in. [558·80 mm.]. The beak is yellow with a red spot at the angle or gonys. [W. P. P.]

2. Distribution.—The breeding-range of this gull includes Jan Mayen Island, Greenland, where it is common, and Arctic America. Dresser gives the following localities: Felix Harbour, Boothia (Ross), Melville Island, Franklin Bay (MacFarlane), Mackenzie Bay (Reed) and Cambridge Bay, Victoria Land (Collinson), and also on the Yukon (Nelson). Saunders, however, thinks that all records from the North Pacific and Bering Sea refer to the glaucous-gull. It is said also to have bred on Novaya Zemlya. (See p. 182.) Its winter range includes Iceland, the northern seas of the British Isles, and the North Sea, while occasionally it strays to the Baltic and the coasts of western France, and has once been recorded from the Adriatic. In North America it has been recorded regularly on Lake Michigan, and casually at Boston ($42\frac{1}{2}°$), Nebraska, and Maryland. [F. C. R. J.]

3. Migration.—A winter visitor of tolerably regular occurrence on the Scottish coasts, but less frequent on the southern and western shores of the British Isles. The numbers that visit us are rather variable: winters in which the birds have been unusually numerous in one part or another are 1872-73, 1874-75, and 1891-92. Most of the visitors are usually immature birds. An Iceland-gull has been recorded from Yorkshire as late as April 18 (cf. Saunders, *Ill. Man. B. B.*, 2nd ed., 1899, p. 681). [A. L. T.]

4. Nest and Eggs.—Does not breed in the British Isles. [F. C. R. J.]

5. Food.—Principally fish, live or dead, according to Naumann (*Vögel Mitteleuropas*, xi. 278). It appears, however, to be, like its congeners, practically omnivorous (Saxby, *Birds of Shetland*, p. 337; Patten, *Aquatic Birds*, p. 444). [F. B. K.]

KITTIWAKE [*Rissa tridactyla* (Linnæus). Sprat-mew (Kent), hacket, hacklet (Devon), kittick (Orkneys), waeg (Shetlands), tarrock (juv.). French, *mouette tridactyle*; German, *Dreizehen-Möve*; Italian, *gabbiano terragnola*].

1. Description.—In the kittiwake the hind-toe is either vestigial or absent, whereby it is distinguished from all other gulls. It may at once be distinguished from the common-gull, with which alone it could be confounded in the field, by the absence of the white "mirror" on the tips of the outermost primaries. When expanded in flight the wing has a conspicuous black triangular tip. The sexes are alike, and there is a fairly well-marked seasonal change of coloration. (Pl. 108.) Length 16 in. [406·40 mm.]. The mantle and wings are of deep bluish slate, the blue extending far up the primaries, save the outermost, which is black along the outer web and for about two inches at the tip: the fourth and fifth quills have white tips and a subterminal band of black, the rest grey: the lower back and tail, like the head, neck, and under parts, are pure white, and the scapulars and secondaries are tipped with white. In winter the occiput is washed with grey, and there is a faint patch of grey in front of the eye and behind the ear-coverts. The orbital rim is red. The beak is greenish yellow with a tinge of red at the base, the inside of the mouth a deep orange-red, the legs and toes black, and the iris is brown. The fledgling and immature birds have a more or less complete band of black or greyish black across the nape, a second broad band across the back of the base of the neck, and black or greyish black marginal and minor coverts, forming a broad band across the outstretched wing. The tail has a broad terminal band of black, and the coverts of the primaries are black: the four outermost primaries are also of a dull slaty black, thereby again differing from those of the adult. The beak and legs are greyish to black, and the iris is dark brown. The nestling in down has the head white, tinged with grey, the back ash-grey, with very faint traces of darker mottlings, and the under parts white. [W. P. P.]

2. **Distribution.**—In England the kittiwake is very local as a breeding species; it is confined to Lundy Island, the coasts of Wales and its adjoining islands, including Anglesey, the Isle of Man, and, on the east side, the Yorkshire cliffs and the Farne Islands. It used to breed in the Scillies, but has not done so since 1900 (*Zoologist*, 1896, p. 344).[1] In Scotland it is more widely distributed, and colonies are found on the east side at St. Abb's Head, the Bass, the Isle of May, and in Kincardine and Aberdeen, as well as in small numbers on the coast of the Moray area. On the west side it is chiefly confined to the islands, where there are innumerable colonies on the Inner and Outer Hebrides, that on the Shiants being probably the largest in our islands. On the northern coast, the Orkneys and Shetlands, it is also very numerous. In Ireland, though very plentiful, it is decidedly local as a breeding species, resorting only to precipitous cliffs and islands. On the east side it only nests on the islands off the

---

[1] Possibly a few pairs still breed in Cornwall near the Land's End.

Co. Dublin coast, but enormous colonies are to be found in Antrim, Donegal, and thence southward to the islands off the Kerry coast. Outside the British Isles this species breeds in great numbers on the Færoes, Iceland, Jan Mayen Island, Franz-Josef Land, Spitzbergen, and Novaya Zemlya, while on the Continent it is said to nest in Brittany, and many colonies exist along the Norwegian coast, especially in the extreme north, and eastward to the Murman coast, and at one or two localities off North Siberia. On the American side it breeds in Greenland and on the coasts of N. America south to the St. Lawrence, but is replaced on the Pacific coast by an allied race. In winter it ranges south to the Canaries, Azores, and Madeira, and has even been recorded from the Cape Verde group and Senegal, as well as from the Mediterranean : while on the American side it migrates to the Great Lakes and the middle Atlantic States, and has occurred in the Bermudas. [F. C. R. J.]

3. Migration.—Resident within our area as a whole, but seasonally absent from various parts. To the south of England it is a winter visitor, but at that season it is scarcely known on many parts of the Irish coasts (cf. Ussher and Warren, B. of Ireland, 1900, p. 346). There appears to be a regular southward movement in autumn with a return journey in March (cf. Ussher and Warren, loc. cit.; etc.), but to what extent these movements may extend beyond our area, both to north and to south, it is impossible to say. In winter the species often congregates in huge flocks (cf. Nelson, B. of Yorks., 1907, p. 691). [A. L. T.]

4. Nest and Eggs.—Generally this species breeds in colonies, sometimes of enormous extent, and always on precipitous cliffs or in the roofs of sea-caves. Almost any little ledge is sufficient to support the nest, which is sometimes only a few feet above the sea, and at other times some hundreds of feet higher. In the great mixed colonies of rock-birds on the British coasts, the Kittiwakes generally occupy the lowest sites, but some of the bird-rocks within the Arctic Circle are occupied almost exclusively by this species from top to bottom. The nest is built chiefly of seaweed and mud, but grasses, moss, and other plants are also used, chiefly for the lining, and both sexes take part in building (see p. 188). (Pl. XLIV.) The eggs vary from two to three in number: in some colonies clutches of two are general, while in others three are quite common. The ground-colour is rather light as a rule, and varies from greyish white to olive-buff, while the markings consist of blotches of brown and ashy shell-marks, sometimes forming a zone. (Pl. J.) Average measurements of 100 eggs, 2·21 × 1·6 in. [56·3 × 40·8 mm·]. Both sexes brood in turn, and, in Iceland, Hantzsch noted that the female incubated chiefly by day and the male by night, changing places about 9 P.M. The kittiwake is a late breeder, and few eggs are laid before the end of May or early in June in the British Isles. Hantzsch estimates the incubation period at 21-24 days, and Naumann at three weeks, but an egg placed in an incubator hatched on the 26th day (W. Evans). Only a single brood is reared during the season. [F. C. R. J.]

5. **Food.**—Principally small fish and their ova, also surface crustaceans (cf. Collett and Nansen, *N. Polar Expedition Scientific Results*, 1900, vol. i. pp. 26, 47; and Collett, *Bird Life in Arctic Norway*). Saxby found fresh-water algæ and some beetles in the stomachs of specimens he shot (*Birds of Shetland*, p. 330). The young are fed on crustaceans and small fish (Naumann, *Vögel Mitteleuropas*, xi. 290) by both parents. [F. B. K.]

The following species and subspecies are described in the supplementary chapter on "Rare Birds":—

Sabine's Gull, *Xema sabini* (Sabine).
Rosy or wedgetailed-gull, *Rhodostethia rosea* (Macgillivray).
Bonaparte's gull, *Larus philadelphia* Ord.
Adriatic gull, *Larus melanocephalus* Temminck.
Great blackheaded-gull, *Larus ichthyäetus* Pallas.
Yellow-legged or Mediterranean herring-gull *Larus argentatus cachinnans*, Pallas.
Ivory gull, *Pagophila eburnea* (Phipps). [F. C. R. J.]

THE BLACKHEADED-GULL

[F. B. Kirkman]

With the exception of the kittiwake, and three rare species that have been recorded as visiting our shores half a dozen times or less, the British gulls all belong to the genus *Larus*. It contains both the largest, the glaucous and great blackbacked, and the smallest, the little-gull. Between these, the better known species arrange themselves, according to a descending scale of size, as follows :—the herring, Iceland, lesser blackbacked, common, and, finally, the blackheaded, the most widely known of all, and the subject of the present chapter.

Strictly speaking, the name of the species should be "brownheaded," for brown is the colour of its hood. The misnomer, no doubt, owes its origin to the fact that, seen at a distance, the hood has the appearance of being black. Less excusable is the scientific name *ridibundus*, bestowed upon the species by Linnæus. I have never heard anything in the notes it utters to justify the appellation, which might far more appropriately have been applied to some of the larger gulls, the lesser blackbacked, for instance, whose loud emphatic *ha! ha! ha!* could not be described otherwise than as a laugh. Custom, in this case, is likely to prove stronger than fact, and I have no doubt that *ridibundus* and "blackheaded" the species v'll remain.

The brown hood is assumed for the breeding season only; from about June to the end of January or early in February the head is white, except for a brown patch or two.[1] It has been stated that the change from the white to the brown is effected by the entrance

---

[1] Several lose the brown hood as early as the middle of June (F. Heatherley, *in litt.*).

Plate 103

Blackheaded-gulls following the plough

By A. W. Seaby

of fresh pigment into the feathers.[1] As long ago, however, as 1866, Mr. H. Blake Knox pointed out in the *Zoologist*[2] that the change is by moult, and that the new brown feathers sprout up from under the white and displace them. He limited the spring moult to the head. Dr. P. H. Bahr has since shown that it extends to the breast and back as well.[3]

Soon after it assumes the brown hood, the blackhead moves towards its summer quarters. Before actually settling on the nesting-ground, it appears to frequent the fields in the neighbourhood for three or more weeks. At Scoulton, in Norfolk, where the birds are "preserved" for the sake of their eggs, and are therefore carefully watched, I was told that the first arrivals do not alight on the nesting-ground till March 17th-20th, but they are seen in numbers in the surrounding country as early as mid-February. Their habits at this period have yet to be closely studied, and should prove full of interest, for it is then, no doubt, that they pair off for the season.

The breeding-places, for the most part, may be classed under one or other of two types. The first may be called the marsh or lake colony. That at Scoulton is an example. Here the nests are built on swampy ground, from which the reeds have been cut, at one end of the large island in the lake. Elsewhere they are built among the water-plants growing on the margin of lakes, and also on tussocks rising from lake or swamp. In the Danube delta Mr. Jourdain found nests floating on water five feet deep, with no support except that occasionally provided by a water-lily leaf or stalk. These colonies are found both inland, and, as at Dungeness, near the sea. They form a remarkable contrast to the second and less usual type, of which Ravenglass and Walney are good examples. At these two places the birds nest in thousands, chiefly on the ridges and sides of arid, sun-scorched sandhills above the beach, sometimes hundreds of yards

---

[1] Yarrell, *History of British Birds*, iii. 603.   [2] 1866, p. 361.
[3] *British Birds* (periodical), iii. 105, where a careful and detailed account of the spring moult is given and illustrated.

from the water. Among very unusual breeding-places may be mentioned that at Twigmoor in Lincolnshire,[1] where several pairs build nests of considerable size in fir-trees, though suitable sites of the ordinary type are available near by. It seems as if here the change of nest-place from ground to tree was the result of individual eccentricity.

New gulleries are not infrequently found, and old ones sometimes deserted. Mr. H. E. Forrest records an instance of a gullery at an unpronounceable place, Llyn Mynyddlod, near Bala, which commenced with two pairs in 1888. In 1889 there were ten, and in 1890 over twenty nests. In 1904 the colony numbered about two thousand. Another colony by a small tarn in Denbighshire began similarly with two pairs in 1900, increased to four or five pairs in 1902, and numbered some sixty birds in 1904.[2] The first pairs were perhaps emigrants from an overcrowded colony. Such overcrowding does appear to occur,[3] and no doubt, at the beginning of the season, leads to much fighting. On this point information is required. The earliest date on which I have personally watched the species was April 6, 1911, at Scoulton, about a week before the first nests were built. The pairs were then evidently in possession of nesting-sites. One, presumably the female, remained on the spot, or near it, the whole day, while the male went far afield for food, which he brought to his mate, departing again shortly afterwards as a rule. Though the birds were aggressive, and there was occasional fighting, I did not notice any very strict insistence on proprietary rights. A bird would often drive away another alighting near it, but in so doing it would, instead of returning, sometimes alight near other pairs, and remain unmolested. Again, birds would rise from their own places and either immediately, or after a short flight over the gullery, alight in another part of it, and, likewise, remain unmolested. This occurred also in the evening, when the number of the birds was trebled by the

---

[1] See "Classified Notes."   [2] *Fauna of North Wales*, p. 379.
[3] Cf. Ussher and Warren, *Birds of Ireland*, p. 320; *British Birds* (magazine), iv. 223.

presence not only of the foraging males, but of a large number of immature birds born in the previous summer, and easily identified by the flecks on the wing, the black rim to the tail, and the absence of a brown hood.[1] These came each evening to roost, and departed in the morning. They distributed themselves all over the gullery, possibly returning each to its birthplace.

The number of the non-breeding gulls is not limited to these immature birds, but to what extent, if at all, non-breeding adults are to be found on the breeding-ground itself I do not know. Some are to be seen on the coast throughout the summer.[2] The most remarkable example is provided by a bird which spent the whole summer by itself on the South Walsham Broad, in Norfolk, where its solitary white figure was a striking object seen flying amid the deep green foliage round the lake.

The gulls that pair have, like other species, a love-display. This I saw at Ravenglass on May 16, when the eggs were already hatching. The performance took place away from the breeding-ground on the mud-flats in front of the village. One of the pair, which I judged from her more restrained demeanour to be the female, kept her beak bent vertically downward. Otherwise her attitude did not depart from the usual. Her mate had his beak bent downward, his tail fanned and somewhat deflected, his wings hanging, and his plumage puffed out. He presented a very queer object indeed. Like the female, he kept moving about with precise and formal little steps. Both held themselves stiffly, and looked as if they were performing a ritual. I did not see them face each other; they walked one after the other bowing, and sometimes the cock turned round so that they bowed in opposite directions. The performance was accompanied by a loud discordant note, uttered by one or possibly both. It was continued for several minutes.

[1] At Scoulton I saw none of these immature birds with a brown hood except one, which had, however, the forehead white. At Ravenglass in mid-May (16th to 21st) there were a few about the gullery in the daytime. These had the hood more or less brown. One of them was actually sitting on eggs, but it was the only one I saw thus engaged.

[2] Nelson, *Birds of Yorkshire*, p. 670.

Two days afterwards I saw a pair near their nest advance side by side, wings half spread, beaks deflected, heads thrown somewhat back, but tail not fanned. At intervals they bowed till the tip of the bill touched or almost touched the ground. They then quickly resumed their normal attitudes. The spirit seemed suddenly to move them, and as suddenly to depart from them.

This display may, for all I know, be frequent in the early part of the season (February to March), but during the period that I have watched the species, namely from the beginning of April to the middle of July, I have only seen it performed on the above occasions, though I have been in or near the breeding-ground for days and hours at a time.[1] The same applies to the cosseting or preening of one bird by another, but throughout the breeding season from April to June, and no doubt to some extent in July, the species is in the habit of performing certain actions which are partly expressive of love or affection, and partly of other emotions. These may properly be classed under the head of the language of gesture. Each has a meaning or meanings as intelligible—to me, indeed, often more intelligible than that intended to be conveyed by the sounds the birds utter. There are at least three of these gestures clearly distinguishable, but any two may follow so closely one upon the other as to be continuous.

The first is as follows. The bird, standing, or moving forward towards another bird, suddenly lowers and retracts the neck, so that it, and sometimes also the beak, point forward. Usually, however, the head and beak are tilted upwards, sometimes almost straight upwards. The wings are kept closed, or partly opened and hanging, or spread so as to form with the back a broad flat expanse. The whole attitude of the bird, in particular the upward tilt of the beak, gives it an air of aggressive truculence calculated to be highly offensive to the party approached. When the attitude is

[1] The dates of my visits to colonies of blackheads are as follows:—*Scoulton*, April 6-13, 1911; *Ravenglass*, May 16-21, 1910, May 10-24, 1909; *Dungeness*, June 5-12, 1908; *Ravenglass*, June 14-17, 1909; *Walney*, June 16-July 13, 1905; *Dungeness*, July 7-14, 1907.

assumed as the bird is swimming on the water, the neck lies flat
along the surface, the beak being tilted up as above described. It is
in some cases difficult to attach any precise meaning to this gesture.
I have, for instance, seen it made by a bird when alone. There can be
no doubt, however, that as a rule it is intended to be a menace, for it
is frequently followed either by the prompt departure of the bird
approached or by a scuffle. It is made by other species of Gull. Mr.
Hudson notes it in the case of the glaucous-gull, and compares it to
the snarl of a dog.[1] It might perhaps even more fitly be compared
to a display of fangs. As in the case of dogs, it is not always meant to
be taken seriously. It is not even always meant to mark hostility.
I have seen it made by one of a pair, the male, and followed by
anything but bellicose proceedings. This may possibly be explained
by the fact that the love of male for female has in it an undoubted
element of aggressiveness, one might almost say ferocity.

In the second posture the neck is held erect, and the beak
usually bent downward and sometimes to one side. The crown
feathers are more or less erected, probably as an effect of the downward inclination of the beak and backward poise of the head, while
the wings are usually held part open and away from the body, the so-
called shoulders (wrists) projecting. The general appearance is that
of lofty disdain, but this may not, and, in fact, certainly does not in
most cases represent the feelings of the bird. The attitude resembles
that assumed by the male in the love-display above described. On
both the two occasions on which I have exact records of its being
used independently of the other two "gestures," it appeared to be a
form of greeting or affection. In one of these cases, one of a pair,
which I took to be the female, was building her nest. Her mate,
who had previously been standing near by, was taking a flight round
the gullery. Suddenly I saw the female strike the erect posture and
also bow. She had evidently recognised her mate in the general crowd
of flying birds, for next instant he alighted. My recollection, aided by

[1] *Land's End*, p. 23.

a somewhat unsatisfactorily worded record in my notebook, is that the erect attitude was assumed independently on other occasions, but in all the remaining fourteen instances which I took the trouble to record exactly at the time of observation, it accompanied the first "gesture" twelve times, and that next to be described once. In the fourteenth instance all the three movements occurred together.

On five of the occasions on which it accompanied the first "gesture," the combined movement was intended to be hostile. In one of these cases a bird, which was swimming on the water, first laid its neck flat on the surface, and tilted up the beak, then shot the neck up, deflected the beak, and part spread the wings. It thus approached another bird, and on nearing it, again lowered the neck, tilted up a beak of derision, and finally made a rush and a vicious lunge, which the menaced party lost no time in avoiding. This attack seemed to be entirely unprovoked, and had the appearance of being merely an outlet for exuberant spirits. On the remaining occasions on which the second or erect accompanied the first or neck-forward-and-beak-up movement, there was no appearance of pugnacity. On two of these occasions the attitudes were struck by birds that were standing by themselves. In the other cases, two or three birds would perform one or other or both of the movements, then merely desist and look about as if nothing had occurred. I have described this set of movements in some detail, because it illustrates the difficulty of interpreting the avian language of gesture, and also suggests that the avian mind is a more complex organ than is generally supposed.

The third attitude is the only one with pretensions to beauty. In its complete form the bird raises the white and grey wings and waves them, spreads the tail into a white fan, and bending down the head, almost, but not quite, touches the ground with the tip of the bill.[1] The tail is not invariably spread or the wings raised, or, when raised, waved. The attitude is much like that of the birds when they are

---

[1] It may, of course, at times touch the ground, but on the one occasion on which I saw the position of the beak quite clearly, it did not.

pulling up water-plants for their nests. It usually either follows or precedes an attack on another bird, and did so on all the occasions on which I recorded it as occurring by itself. This was in April. Once in April and once in June I saw it used conjointly with the second or erect attitude. In the first case a pair stood side by side with the beaks nearly touching the ground, and then suddenly and simultaneously raised their necks and deflected their beaks. In the next, a pair began with the neck-forward-and-beak-up movement, passed to the erect posture, and one ended with the third or beak-to-ground posture. In neither case was any hostility apparent. In no case except this did I see the third movement used with the first. As it usually precedes or follows an attack, it might be thought that, like the first, it is a manifestation of pugnacity. But there is nothing in it of the direct menace conveyed by the aggressive forward inclination of the neck and the insolent uptilted beak. Its real significance has yet to be ascertained. It is worth adding that, though I watched carefully, I did not see either this gesture or the other two performed by any of the immature birds which crowded into the Scoulton gullery in the evening.

It must not be supposed that an attack by one gull upon another is necessarily accompanied by gestures. In most cases, when a bird has made up its mind to drive home the assault, it does so without preliminaries. If the offender thinks it worth while to resist, both birds either rise in the air fluttering, in which case I have not observed that anything results but noise, or they scuffle and sprawl on the ground. In the latter case there is little to be seen but a confusion of white wings. On the only occasion on which I had a clear view, one of the combatants seized the other by the bill, and they continued to tug and tumble for a few seconds till they had had enough. These squabbles seemed to arise from the proximity of one bird to the nest of another. Quite distinct from them are the more or less mock combats in the air, marked chiefly by pretty upward sweeps and curves, the wings outspread and motionless, or nearly so, red legs dangling, red mouth wide open and vociferous, followed by downward swoops. Distinct again are the assaults of flying birds upon sitting. The best time for watching these is during the periodical simultaneous flights that a number of birds in a colony or part of a colony will make when recently disturbed. The upflight from the ground is comparatively silent. Even when not looking at the part of the gullery in which the upflight occurs, one knows it is taking place by the sudden hush. No sooner do the birds halt in mid-air than the noise again begins, and it goes on increasing on the down-flight till it reaches a climax at the moment that precedes alighting. Then there is a wild confusion of white fluttering wings, of red legs flashing in the sun, and from the red wide open beaks there issues a fast and furious storm of screams. It is at this moment that sitting birds have to be on their guard; they often sit, with neck and head upstretched and beak wide open, to defend themselves from a playful dig in the back from their over-excited mates or comrades in the air.

In the universal incessant chorus of disapproval that descends upon the intruder into a colony of blackheads, it is no easy task to distinguish the different notes that are uttered, especially when allowance is made for distance and differences in pitch. One note, however, it is possible to study in its most intimate details, and that is the harsh screech that ends the swoop of the angry gulls near whose nest one happens to be. It is uttered at the moment when the bird, instead of dashing itself upon the object of its wrath, as it seems about to do, suddenly checks its flight and swerves away to mount and repeat the attack. It is repeated again and again till the victim is heartily glad to beat a retreat. It provides the strong notes in the gull orchestra, and may be variously syllabled as a strident *qwerrr*, *qworrr*, or *qwarrrr*. During the descent to the attack the bird utters a rapid *tt, tt, trrr, tt, tt, trrr*. This note forms a background of sound in the general chorus. A third note is a curious *ttuk*, or *ttup!* which may be uttered as a single sound, or repeated, sometimes slowly, sometimes rapidly. When rapid, it seems to merge into the *tt, tt, trrr,* but of this I am not certain. It is often interspersed with *qwarrr* sound, and appears to be a general note of alarm, possibly a call-note. It may be heard punctuating the general chorus. It is probably this note that is supposed to resemble a laugh. Another note familiar to those who have watched blackheads in winter, whether from the Thames Embankment or on the coast, is a petulant little infantile scream, uttered usually when the birds are bickering. I have never heard it on the breeding-ground.

The boldness of the blackheads when they have eggs or young is in strong contrast to their shyness at the beginning of the season, when an incautious approach to the gullery will send them off into the air for an hour or more. Such at least was my experience at Scoulton early in April (1911). At this period the hens, for such I supposed them to be, were, as already noted, in occupation of the sites of their future nests, and there they appeared to remain throughout the day, waiting apparently for the first impulses of the nest-building instinct. The cocks busied themselves in searching for food in the neighbouring fields. As they often went several miles for this, there is nothing astonishing in the fact that they did not, after the manner of Thrushes, bring back a mere beakful of worms or insects. They brought back a large supply, and carried it inside. On reaching the gullery they alighted near their mates, and after some persuasive cosseting of the neck and beak, were induced to disburden. Most of the meal was pecked up from the ground by the hen, but the cock often fed with her side by side.

According to Naumann, the nest is built by both sexes. At Scoulton I had time to see only two instances of nest-building, the earliest (April 12-13). In both cases the work was done by one bird, presumably the hen. In the first instance the male was absent, in the second he was present, but took no interest in the proceedings. As the cock is absent most of the time in the fields, it is difficult to believe that he can take much part in building, if he takes any. The process of building was simple. The hen pulled water-plants, carried them to the nest-spot, dropped them, and occasionally sat down upon them and worked them into shape.

The material of the nest varies of course with the surroundings. Among the sandhills of Ravenglass and Walney they are mostly made of marram grass, and are of all sizes. The largest nest I have ever seen was at Ravenglass. It was made of the rough dry stalks of some large plant, to which was added a lining of marram grass. It was nearly two feet high, and the material would have filled two buckets. (Pl. xliv.) Large structures made of water-plants are also found in marsh or lake colonies.

The species seems able to modify very readily its nest-building habits. At Dungeness, for instance, in 1908, a few pairs, apparently unable to find place for their nests in the main colony among the reeds, built them on dry ground near by, and, instead of water-plants, used moss, lichen, dry grass and twigs, and were content with much smaller nests than those of their fellows.

Laying begins in April, the time varying with season and latitude. Incubation is said to commence after the first egg is laid.[1] Both sexes share in the task. On one occasion I saw a bird alight and push its mate off the nest with its breast, so eager was it to settle on the eggs. Dr. F. Heatherley informs me that he saw pairs change places every half-hour. This happened many times and with great regularity.

The chicks are hatched on successive days.[1] They are pretty little down-covered creatures, and are soon able to move about. They quit the nest when alarmed, and, in the sandhill colonies of Ravenglass and Walney, conceal themselves in the marram grass. I have seen one led back to the nest by the parent bird, who achieved her purpose by calling to it and making snuggling movements with her breast. They return also of their own accord. I have had no opportunity of watching their behaviour in a marsh colony, except to note that they were swimming when about a week old. Dr. F. Heatherley informs

[1] *British Birds* (Mag.), iv. 141 (E. B. Dunlop).

me he has seen them at Llandegla, N. Wales, climb of their own accord in and out of the nests into the water. On being alarmed they usually left the nests, and hid in their foundations. According to Dr. P. H. Bahr,[1] the instinct of the chicks, when four to five days old, is to take to the water on the approach of danger, often with fatal results, for some are snapped up by the larger species of Gull, and others, when they become soaked with water, develop a curious sort of cramp from which they do not recover. Mr. H. A. Macpherson, on the other hand, observed that though the chicks are able to swim strongly at an early age, they rarely attempt to escape capture by taking to the water, but usually run to the bank and hide in the sedges or rushes.[2] This apparent contradiction is no doubt to be explained by some difference in the local circumstances, but lack of cover was not one of them, for the island on which nested the colony referred to by Dr. Bahr was "remarkable for its excessive verdure, due to the guano deposited by generations of these gulls."

As the young feather, they become much more inclined to wander away. At Walney and Ravenglass they tend to quit the sand-ridges for the flats, where they are mercilessly attacked and often killed by the terns, as already described (p. 91). Both chicks and fledged young are frequently maltreated by adults of their own species, sometimes for no apparent reason, and at other times because they are trespassing. Many die either from disease or starvation. At Walney, in 1905, I saw the ground strewn with the dead bodies of fledged young, and also a fair number of old birds.[3]

At all ages the young are fed by the parents in the same way as the cock feeds the hen, that is, the food is disgorged and is either picked up by the young off the ground, or snatched before it reaches the ground, sometimes even before it is out of the old bird's beak.

On one occasion Dr. Heatherley saw the parent bird bring up a

[1] *Home Life of Marsh Birds*, p. 56.
[2] *Fauna of Lakeland*, p. 425.
[3] Cf. Macpherson, *Fauna of Lakeland*, p. 425; *British Birds* (Mag.), iii. 201.

mass the size of a walnut, and keep it between the mandibles for the young to peck at.[1]

The young are able to fly well in July, and towards the end of this month both old and young quit the gullery. The species is to be seen round our coasts in winter, also inland, where it is in parts almost as familiar an object in the fields as the rooks, and like them, sometimes with them, is to be seen following the plough. Its migratory movements are somewhat confusing, and have yet to be clearly made out. What is known of them is summarised in the "Classified Notes."

It is worth noting that when flying to roost the species has been seen to adopt the chevron or V-shaped formation. This was observed by Mr. T. A. Coward in the case of the birds that roost at Rostherne Mere, in Cheshire, where they are to be seen in very large numbers every evening from November to February.[2]

Full particulars of the food of this omnivorous species will be found in the "Classified Notes." It obtains it not only on land and water but also in the air. Often in the twilight of a summer evening the white forms of the birds may be seen flitting here and there in pursuit of moths and other insects.[3]

Two peculiarities in its feeding habits call for special notice. One has been called "pool dancing" by Professor Patten, and is thus described:—"It is this: a blackheaded-gull wades into a little pool, the water of which is only deep enough to cover part of its feet; it then lowers it head and looks at the bottom. Finding no food, it at once commences to prance up and down, stirring up the sediment, out of which it picks various marine creatures and fragments of seaweed. I have seen many of these birds at this performance along the mud-flats of Dublin Bay."[4] Mr. A. H. Patterson has noticed the same several times on the Breydon mud-flats (Norfolk), and compares it to the dancing of a hornpipe.[5]

[1] In litt. [2] Fauna of Cheshire, i. 428. Cf. also Naumann, Vögel Mitteleuropas, xi. 212.
[3] Ussher and Warren, Birds of Ireland, p. 330; Zoologist, 1843, 246; 1844, 455, 577-78; 1002, 216.
[4] Patten, Aquatic Birds, p. 414.
[5] Notes of an East Coast Naturalist, p. 142.

The other feeding habit calling for mention may properly be called parasitic. It takes the form of keeping close attendance upon other species, and of attempting to snatch from them the food they draw up from the ground. I have repeatedly seen blackheads, two or three at a time, standing round an oystercatcher, watching with interest the disappearance of his long red beak into the mud, and with still greater interest the emergence of the same. The seapie was under no illusions as to the meaning of these attentions, and, when it drew a worm up, lost no time in making off with its prey. This was in May at Raven-glass. Plovers are pestered in the same way, especially in hard weather during winter, and apparently with some success.

Having regard to these facts, one is surely justified in feeling that the sense attached to the word "gull" in the dictionaries is no longer appropriate.

THE LITTLE-GULL

[F. C. R. JOURDAIN]

In general appearance this gull is a miniature edition of the blackheaded-gull, and even more closely resembles the Adriatic or Mediterranean blackheaded-gull on a small scale. It is a tolerably regular visitor in small numbers to the British Isles, occurring chiefly on our east coast, and sometimes in considerable numbers in autumn and winter. The most important visitations of which we have records are those of 1866 and 1868 to East Yorkshire, while in the winter of 1869 and February 1870 still larger numbers were met with along the whole of the east coast of England. Northward it has been recorded from the Shetlands, along the Channel westward to the Scillies, occasionally along the west side of Great Britain, and very rarely in Ireland. As, however, it is clear that this species is extending its breeding-grounds westward, it is probable that it will occur more frequently on our coasts in future years. It is still mainly an East European and Asiatic form at the present time, but during the winter may be met with in fair numbers even in the western Mediterranean. Amongst its favourite winter haunts may be mentioned the lagoons of Tunisia, where it is common, and, as noticed almost everywhere, very tame and confiding in its habits. Many birds seem not to cross the Mediterranean, but to winter round the coasts of Greece and the adjacent islands. Lord Lilford gives the date of their departure from the Ionian Islands as about the beginning of March, but in the Black Sea they seem to remain much later, for Mr. W. H. Simpson (*Ibis*, 1861, p. 362) describes immense numbers as haunting the lagoons north of Kustendji (Constanța) between April 20th and 24th. "The flocks of *Larus minutus*, associated with a few individuals of *Sterna cantiaca*, were literally swarming in the air a few feet above the surface of the water, like swallows over a river on a summer's evening. Far as the eye could reach, looking northwards down the lake, these elegant little birds were to be seen on the feed, dashing to and fro most actively. . . . A few days later and the thousands have become hundreds; yet a few days more and these will have dwindled down to tens, so that by the middle of May it is possible that not a pair will remain behind." Probably Simpson's supposition is correct, for when we visited the lagoons of the eastern Dobrodjea in the latter part of May 1911, we saw no trace of this species, though its larger relative, the Mediterranean blackheaded-gull, was present in thousands, and just beginning to breed. There is, however, some slight evidence that a pair or two occasionally stay to breed in this region, as so many other northern forms are said to do. Dr. W. H. Cullen sent a skin of this species to Professor Newton, stated to have been shot from a nest with a single egg, on June 5, 1866.[1] R. von Dombrowski also includes it in his list of Roumanian birds as breeding sparingly on Lakes Razim and Sinöe. Confirmation of these records by independent testimony is, however, very desirable. In South Russia the

[1] *Ootheca Wolleyana*, ii. p. 314.

evidence of its breeding is not satisfactory, and von Nordmann's statement that it nests in the salt lakes in numbers is not accepted by recent
writers. S. A. Buturlin only mentions the Moscow, Kazan, Ufa, and
Perm governments as Russian breeding-places, in addition to the
better-known colonies in the Baltic Provinces, near Archangel and on
Lake Ladoga. Here, however, we are on firmer ground: Blasius,
Lilljeborg, and Meves have all visited and described the Ladoga
colonies, while V. Russow met with three breeding-places in Esthonia,
Stoll found a large colony in Livonia, and in 1887 J. A. Sandman
found considerable numbers breeding on an islet near Karlö in the
Gulf of Bothnia. From here we can trace its spread westward to
Gotland and East Sweden (S. Upland and Jemtland, 1901), while in
N. Germany Henrici found it breeding on the Drausen-see in 1899
(where von Homeyer had obtained a breeding pair in 1847), and it
was recorded as nesting at Rossitten in 1902. Its western limits are,
however, the islands in the Danish fjords. At Rödby, in Laaland,
there seems little doubt that it bred in 1901; and at Klaegbanken, in
the Ringkjöbing Fjord, nests were found in 1904, and it had probably
been breeding there for several years previously. Even as far back as
the spring of 1893 Chapman saw a couple of pairs only a few miles away
from here. The whole of this district is now strictly preserved by the
Danish Government, and can only be visited by special permission.
Klaegbanken itself is a low flat island of mud and sand, varying in
size according to the height of the water in the fjord, and covered
with coarse herbage. A strong south-west wind lasting for a day or
two will drive the waters of the North Sea up the narrow entrance of
the fjord and materially raise the level of the water inside, till a change
in the wind allows the water to flow out again. Here, in this desolate
region, with nothing to break the horizon but ghostly white houses
and clumps of trees, with the tell-tale shimmering streak of light
beneath them which shows that they are due to mirage, lie the breeding-haunts of thousands of sandwich-terns, blackheaded-gulls, common
and Arctic terns, together with a few little-gulls. A few miles away

lies the grassy peninsula of Tipper, where avocets, dunlin, redshanks, ruffs, and a few godwit breed. To give some idea of the numbers of birds breeding in this sanctuary, we may state that in 1900 Rambusch estimated the number of blackheaded-gulls at about 10,000, Sandwich-terns at 4000, while there were also large colonies of gullbilled-terns, as well as common and Arctic-terns, on this one island alone. On landing, careful observation results in the discovery of about ten pairs of little-gulls among the whirling crowd of snowy forms over-head. There is not the slightest difficulty in distinguishing them on the wing when once seen: in full breeding plumage the black head and small size, as well as the white tipped primaries, render them unmistakable. Their notes, too, are very characteristic. Henrici compares the cry to that of the redshank, though the comparison seems rather fanciful, and writes it as "*kei keikei-keikei-keikei-keikei*,[1] etc., often repeated twenty to forty times in succession, so that when several birds are calling together, the result is quite a concert in the air. They are most noisy when dashing about in small flocks close to the ground or playing about at a great height in the air in fine weather, and the penetrating sound can be heard above the deafen-ing babel caused by the creaking notes of the terns and the deeper cries of the blackheaded-gulls. Other notes described by the same writer are a low "*tok, tok, tok*," while resting, or a softly uttered "*kie, kie*" (or *ke, ke, ke*) in flight. They are sociable birds, and in early spring are generally to be seen in small flocks. Sometimes they ascend to such a height that they are scarcely visible, and only the frequently uttered call-note enables us to detect their presence.

The nests must be sought for in the wettest part of the island. Here on the 'slob-land,' overgrown with clumps of rushes, may be found the nests, which look like miniature blackheaded-gulls' nests. On Klaegbanken Christensen noticed that they were built of coarse stalks, which must have been brought by the birds from some distance.

---

[1] Pronounced in English *kayeé, kayeé, kayeé*, etc.

Sandman and Henrici found the nests floating on the surface of the water or partly supported by the growing clumps of *Phragmites* and water-aloe (*Stratiotes*), while the principal materials used were the stalks of *Scirpus lacustris*. Owing to the situation of the nests, a very slight rise in the water-level is enough to submerge them, and this has been observed to happen both in Denmark and also in Esthonia. Russow states that on visiting a colony after a disaster of this kind about a week later, he found the birds busily building much more substantial structures, which had a foundation about six inches deep.[1] He also noticed that in Esthonia many nests were built on the masses of dead reeds crushed down by snow and ice in the winter.

In the artless nest the eggs, usually two, or more typically three, in number, are laid in the latter half of May and early in June. Christensen found clutches highly incubated in Denmark on May 15, 1904, but on Karlö full clutches were first found by Sandman between June 7th and 10th. Russow says that four eggs have exceptionally been found in one nest.[2] They bear a strong resemblance to those of the common-tern, but do not show the traces of "run" colour, due to rotation in the oviduct, and are a rather fuller and more rounded oval in shape, though distinctly pointed at the small end. There is generally also a tendency to an oily gloss, which is absent from the tern's eggs, and the greenish olive ground is also characteristic of the gull's eggs. As the two species are very often found breeding together, careful identification is always necessary. Exceptionally, blue eggs have occurred, as in so many other species of gulls.

The average size of 97 eggs is 1·63 × 1·18 in. [41·5 × 30·1 mm.]. When incubation has begun the birds cease to fly up and down the lake with continuous cries. At daybreak the non-incubating birds go away to feed, and do not return till nearly midday, when they are met by their mates, who change places with them. We still lack information

---

[1] *Die Ornis Ehst-, Liv- und Curland's*, p. 194.
[2] One instance of five eggs in a nest has been recorded (W. Meves).

as to the duration of the incubation period. When the young are
hatched they skulk about among the thick growth of water-plants like
the young blackheaded-gulls.

W. Meves, in his account of the Ladoga colonies contributed to
Dresser's *Birds of Europe*, states that the noisy flocks which surround
the visitor to a breeding colony from time to time all leave as if by
some common impulse, but in a short time return and again renew
their expostulations. He also observed young of the previous year,
easily distinguished by the black band on the tail and the light head,
but they did not appear to be breeding.

A great part of the food of this species consists of insects captured
on the wing. It also takes small fish, and at Lake Ladoga Meves
found that these constituted its chief food. They are picked up from
the surface of the water, and the gulls do not drop into it in the way
that the terns do, so that it is rare, according to Naumann, to see
more than the bill, head, and neck immersed in the water. The same
writer includes larvæ of dragon-flies, water-beetles, and molluscs in its
dietary. The huge swarms of gnats and midges, which are hatched off
in millions from the waters of these shallow lakes, also provide a great
store of food, especially *Chironomus riparius*, according to Christoleit.
Towards evening great columns of these insects appear, dancing
in the air, and the gulls dash through and through them with
rapid flight, seldom rising more than twenty feet above the water,
which is about the height to which the flies ascend. About the
middle of July the young are fully fledged and are strong on the
wing. Soon afterwards they begin to leave their breeding-haunts,
and by the beginning of August all have disappeared for their
winter quarters.

THE COMMON, HERRING, AND BLACKBACKED-GULLS

[F. B. Kirkman]

The four gulls that form the subject of this chapter provide a convenient group for purposes of comparison. The comparison cannot at present be complete, for our knowledge of the domestic details of their lives is defective. But the purpose of this book is still served, if we show what has to be found out as well as what has been found out, and so clear the way for future work.

In point of size and coloration the greater blackbacked-gull is to the lesser blackbacked as the herring-gull is to the common. The first two measure in length about 29 and 22 inches, the second two 24 and 18 inches respectively. The first two have blackish or slate-coloured backs and wings, the other two grey. In adult plumage the larger of the first two, the great blackbacked, has flesh-coloured legs, so has the larger of the second two, the herring-gull. The smaller species in both cases have yellowish legs. The common-gull alone of the four lacks the red blotch on the downward projecting angle of the lower mandible. The young of all four species have a curious mottled plumage, of which the dominant colour is brown, and it is only after three to five years of gradual transformation that they assume the complete garb of maturity. Except in size and shape, they look entirely different from the adults, and no doubt illustrate a phase in the evolution of the species from the ancestral plover-like form.

The distribution of the species in the British Isles presents many points of interest. Why, for instance, does the common-gull limit its breeding-range to Scotland and the west of Ireland? There are many suitable places in England and Wales. Why, again, is the herring-gull the most widely distributed, not only in but outside the

British Isles, one form or another of it being found in most parts of the northern hemisphere? No answer to the first suggests itself at present. The probable answer to the second is that the success of the herring-gull is due to the fact that, for some reason, it has proved the most adaptable in its nesting-habits. It not only nests where its congeners nest, on tops of islands and cliffs, on moor or shore, but also on the face of precipitous cliffs, which they avoid with the exception of an occasional lesser blackbacked-gull.

All four are gregarious in the breeding season, but single pairs may be found by themselves, or associated with another species. Solitary single pairs of breeding common-gulls are not infrequently seen on the Scottish hills.[1] Single pairs of greater blackbacks are likewise not uncommon. A notable instance is provided by a pair which elected to nest on a grassy island in Lough Conn (N. Mayo) amid a large colony of blackheaded-gulls. To the latter the honour proved to be anything but agreeable, for no small number of their eggs went to satisfy the appetite of their two gigantic relatives.[2]

On the interesting subject of displays or gestures very little information is available. Herring-gulls in captivity have been seen bowing to each other, accompanying the performance with a note, not described. The male selected his mate from three females, and it is stated that the chosen one persecuted her two companions, prompted, perhaps, by some primitive sense of jealousy.[3] Mr. E. Selous, in a chapter on the blackbacked and herring-gulls, observes that "Gulls have no very salient or pronounced courting antics—I mean I have observed none,—and, in the same sense, there is no special display of the plumage by one sex to the other. When amorous, they walk about closely together, stopping at intervals and standing face to face. Then lowering their heads, they bring their bills into contact, either just touching or drawing them once or twice across each other, or else grasping with and interlocking them like pigeons, raising them a little,

[1] Harvie-Brown, *Fauna of the Moray Basin*, p. 215; ibid., *Fauna of Argyle and the Inner Hebrides*, p. 190.
[2] *Zoologist*, 1911, 349.   [3] *Irish Naturalist*, 1898, 255 (Wm. M'Endoo).

Plate 104

Common-gulls in winter

By Winifred Austen

and again depressing the heads with them thus united as do they. After this they toss up their heads into the air, and open and close their beaks once or twice in a manner almost too soft to be called a snap. Sometimes they will just drop their heads and raise them again quickly, without making much action with the bills. This is dalliance, and between each little bout of it the two will make little fidgety, more-awaiting steps, close about one another. Always, however, or almost always, one of the birds—and this one I take to be the female —is more eager, has a more soliciting manner and tender begging look than the other. It is she who, as a rule, commences and draws the male bird on. She looks fondly up at him, and raising her bill to his, as though beseeching a kiss, just touches with it, in raising, the feathers of his throat—an action light but full of endearment. And in every way she shows herself the most desirous, and, in fact, so worries and pesters the poor male gull that often, to avoid her importunities, he flies away." [1] The behaviour of the female described in the last few lines is not unlike that of the blackheaded-gull when trying to persuade her mate to produce from his inside the meal she expects ; and the male blackhead, when pestered, often wears just the same worried look. These fondling movements of the birds, watched by Mr. Selous, if not immediately prompted by desire for food, may have had their origin in the same. An action first used for one purpose may well come to be used for another quite different.

A habit frequently observed among the larger gulls, including the blackbacked and herring, if not the common species, is that of suddenly stretching forward the neck, in a somewhat regurgitative manner, opening wide the beak, and uttering a wild exultant scream. Also no doubt common to all is the menacing gesture, already described in the case of the blackhead (p. 140), which takes the form of lowering the neck and head, and giving to the beak an insulting upward tilt. I have seen it made by an immature herring-gull in St. James's Park.

[1] *Bird Watching*, p. 111.

Though gulls, as already noted in the case of the blackhead, frequently limit the expression of their pugnacity to mere menace, serious fights do occur not infrequently. Mr. Selous, in the chapter above mentioned, gives instances that show something of the method of combat. In one instance the combatants were two blackbacks (species not stated), who fought beneath a living canopy formed by a whirling crowd of interested fellow-gulls. One stood on the defensive, "beating or trying to beat off with wings and beak the continual eager rushes of his assailant. Many times they closed and went struggling and flapping over the ground, attended all the time by gulls in the air and gulls walking about or near them." The bird on the defensive "finally beat off its assailant, who now took to the air. Sweeping backwards and forwards above the hated one, it made each time that it passed a little drop down upon it with dangling legs, and delivered, or tried to deliver, a blow with the feet, a strategy which the other met by springing up and striking with the beak."[1] The close interest taken by their fellows in the fight illustrates a fact frequently observed in bird-life. This interest is by no means confined to individuals of the same species as the combatants. I have observed a duel between two robins followed with the closest attention by a hen-sparrow, to whom the proceeding appeared to afford unbounded entertainment, judging from her loud and excited chuckles. Many such instances might be quoted.[2]

In another instance the opponents, two herring-gulls, fought in the manner already noted in the case of the blackheaded species (p. 143). Each seized the other by the beak, and tugged, the one or the other being dragged about, a process which it sought to resist by spreading its wings and exerting pressure with them upon the ground.

[1] *Bird Watching*, p. 107.
[2] The interest taken by birds in the quarrels of their fellows may be the possible explanation of a curious incident related by C. Rubow in his *Life of the Common-gull* (pages unnumbered). He states that he saw one of a colony of common-gulls "sentenced to death and executed by its comrades." It was certainly killed, but only one wound was found upon the body, and that "a deadly one in the back of the neck." It was, therefore, apparently attacked by one bird only. But whatever be the explanation, the view that the bird's death was due to a judicial execution is not proved by the evidence given.

The struggle lasted several minutes, and was resumed the instant the beaks were unlocked. " At length there was a very violent struggle, and the bird that seemed to have the advantage in its hold, by advancing upon the other while never-relaxing this, forced its head backwards and at length right down upon its back, the bird so treated being obviously much distressed. At last, with a violent effort, this latter got its bill free, and the two, grappling together, and one, now, seizing hold of the other's wing, rolled together down the steep face of the rock. At the bottom they separated." Later the combat was resumed, and ended in one bird flying off and alighting, to be pursued by the other and attacked " with savage sweeps from side to side." [1]

The cause of these contests is not always clear. They appear to be sometimes due to jealousy, sometimes to the presence of one bird near the nest or nesting-place of another, and certainly at times to the objection shown by one to the collection of nest material by another in its immediate proximity.[2]

On turning to the sounds uttered by these species, one is met by the usual hindrance to exact comparison arising from the absence of a scientific phonetic notation. The same sounds are variously syllabled by different observers, and it is at times only possible to recognise their identity from circumstantial evidence. The same sounds are of course sometimes heard differently by different ears, and in this case a scientific notation would be of no avail. But there can be no doubt that, if we had a separate symbol to represent each separate avian sound,—a system corresponding to that in use among modern language scholars and teachers,—not only would the work of comparison be facilitated, but the study of the sounds would become far more careful and exact.

Notwithstanding the variety of syllabic characters used, it is possible to detect a strong family likeness in the notes uttered by the present species. Their alarm-notes have all been variously described as hoarse cackles, laughs, and even barks. That of the common has been figured as *kak!* *skak!* or *yak!* and is the most distinctive.

[1] *Bird Watching*, p. 108.   [2] *Ibid.*, p. 104.

Those of the herring and blackbacked-gulls resemble each other closely, though by no means so closely that they cannot be differentiated. They have all been syllabled as *ha! ha! ha!* or *ha! ha! ha! hà!* also in the case of the lesser blackbacked and herring as a repetition of *ags* or *cags* and the like, and in the case of the greater blackback as *og!* or *ugh!* Heard closely, the laugh, or *ha! ha! ha!* resolves itself for me, in the case of the herring and lesser blackback, into *quow!* or *ow!* Uttered by the latter species, it sounds somewhat like the gasp made by a person who has received a sudden blow in the wind. It is sometimes slightly dissyllabic, and may perhaps be described as sharper and more explosive than the corresponding note of the herring-gull. When excited, both species utter the notes in rapid succession, and may bring them to a climax, when at close quarters, with a loud harsh scream. Usually the *ow!* or *quow!* is uttered three or four times. The corresponding note of the great blackback is described by Lord Lilford as "a sharp hoarse bark, and has a peculiar character about it that distinguishes this from all other British gulls."[1] That these notes are used only to indicate alarm is improbable. More information as to specific occasions on which they are heard has yet to be collected.

Considerable ingenuity has been exercised in figuring what is, with the note just described, the chief utterance of the four species. It is heard usually when the birds are on the wing, and is perhaps a call-note, and an expression of well-being. The most exact of the attempts to represent that of the common-gull is by a German ornithologist, J. Rohweder, and it illustrates admirably the difficulty attendant on the use of the ordinary or nomic spelling. The rendering is *gnjiiäh*, a high pitched nasalised note, with but slight stress on the last syllable.[2] I venture to believe that to those who have not heard the sound itself, the symbol is likely to give but a dim idea of the exact rendering, and it will, no doubt, be differently

---

[1] *Birds of Northamptonshire*, ii. 259.
[2] Naumann, *Vögel Mitteleuropas*, xi. 228: "hoch und wie einer Art nasaler Dämpfung *gnjiiah*."

rendered by different individuals. Other renderings are *skiah!* and *kljah* or *klijrrah*, also both by Germans, Naumann and Droste-Hulshoff.

The corresponding note of the herring and lesser blackbacked has been figured as *kiow*. This, or *quee-ow*, is my own rendering. The sound will be better recognised by the description of it as a plaintive mewing note, but with something free and wild about it that is not reminiscent of the domestic cat upon the hearth. It is one of those singular sounds that seems to voice the spirit of the place in which one hears it often uttered, the mid-air breezy planes above tall cliffs or rocky isles between the blues of sea and sky.

The rendering of the kindred note of the great blackback is *kgau* or *kjauvihs*, neither of which convey much.

Other notes are uttered by the species which have yet to be closely studied. I have heard from the lesser blackbacked a note something like *or!* or *yor!* It was uttered by one of a pair while its mate sounded the usual explosive *ow!* When I drew nearer their young, both uttered an excited *ow!* rapidly repeated. The same species has a lower note, with which the *ow!* is sometimes intermixed, like *quk, quk, quk*. This, intensified, is the exultant scream flung forth by the birds when suddenly they stretch out their necks in the re-gurgitative manner aforesaid. The scream begins often with a sound like *oo-wèrrr*; then follow the *quks*, and occasionally a few cackling *ows!* Corresponding to the *yor!* of the blackback is the *ee-horr* or *quee-hor* of the herring, which I heard uttered by a few birds when sailing, while their fellows uttered the more usual *ows!* I have not noted the sounds composing the scream of this species. Naumann makes it a repetition of its ordinary note, the *kiow* or *kjow*. According to A. von Homeyer, the bellow of the great blackback is its alarm-note, a rapid volley of loud *ogs*.[1] From this it appears that the two species last-mentioned use different kinds of notes in their scream, which is, on the face of it, somewhat impro-

---

[1] Naumann, *Vögel Mitteleuropas*, xi. 263.

bable. It may be that, like the lesser blackback, they introduce into it more than one kind of note.

Little is recorded of the way in which these gulls build their nests. Herring-gulls have been seen to use their feet to make preliminary scrapes, and their breasts to shape the nest.[1] A lesser blackback kept in captivity by Saxby used regularly to scrape out a rude sort of nest, and sit perseveringly in it for weeks. But eggs were of greater interest to her as articles of diet than for purposes of incubation, and if any were introduced while she was absent, she would, on her return, break them open with her bill and swallow most of the contents.[2]

The material of the nest, which is dropped on to the chosen spot and shaped into an untidy couch, varies from place to place, according to what is available in the neighbourhood. Among the kinds of material used are twigs, seaweed, grass, heather, sea-campion, rushes, feathers, moss, and thrift. The amount of material used varies greatly from nest to nest. Some birds content themselves with the scantiest lining, others make bulky structures. Lord Lilford noted that nests of the, great blackbacked seen by him on the Scillies were accumulations of several years, and were composed of piles of withered marram and other grasses with a few sticks and some fragments of rabbit-skin.[3]

There is, as in the case of the blackheaded species, some evidence to show that the work of nest-building is performed by the female only, but nothing definite can be said.[4] Laying begins in May. The male common-gull shares in incubation,[5] and as the males of the other species are provided with brooding spots, it may be assumed that they do likewise.

The chicks appear to quit the nest almost as soon as they can toddle away. According to the observations of Mr. C. J. King

---

[1] O. Leege, *Vögel der friesischen Inseln*, 1905; Saxby, *Birds of Shetland*, p. 340.
[2] *Ibid.*, p. 337.   [3] *Birds of Northamptonshire*, ii. 259.
[4] E. Selous, *Bird Watching*, p. 104.
[5] C. Rubow, *Life of the Common-gull* (pages unnumbered).

in the Scilly Isles, they do this habitually, whether alarmed by the approach of intruders or not.[1] They are found hiding among the stones, in the grass, or under any suitable cover. Intruders are attacked by the parent birds after the manner adopted by the blackheaded-gulls. They swoop down one after the other, and when near the enemy utter a loud and discordant scream. But they do not appear to be, as a rule, so persistent in their efforts as the smaller species.

Herring and common-gulls have been observed to feed their young by regurgitation, in the same way as the blackhead.[2] It is asserted that the common-gull will swallow again a morsel, such as a mouse or mole, which proves too large for the young. If after this second process of digestion it is still too large, then once more it is swallowed, and in due course regurgitated.[3] In the case of a bird observed by Mr. Selous, the food, a small fish apparently, was disgorged on to the ground, and pushed with the beak towards the chick, who thereupon swallowed it. Dr. F. Heatherley has seen the greater blackbacked regurgitate food into the beak of its young.[4] I can find no specific record of how the lesser blackbacked performs the act of feeding its offspring, but have no doubt whatever that it resembles its congeners. A list of the kinds of food usually given to the young will be found in the " Classified Notes."

The food of the parents is exceedingly varied, more so even than that of the Crow tribe, for it not only includes everything eaten by the latter, but also much in addition drawn from the open sea. All four species may be seen in large numbers following shoals of surface-swimming fish. They do not plunge into the water after the manner of Terns, but descend in a curve, and immerse, as a rule, only the head and beak. They are not above snatching fish from the nets and from one another. According to Mr. E. Selous' observations, a bird which

[1] But according to C. Rubow (op. cit.), the chicks of the common-gull find their way back to the nest. This is also undoubtedly the case with the blackheaded-gull (see p. 146).
[2] Naumann, Vögel Mitteleuropas, xi. 252; E. Selous, Bird Watching, p. 119; C. J. King, in litt.; and for the common-gull, Rubow, Life of the Common-gull (pages unnumbered).
[3] Rubow, op. cit.　　　　[4] In litt.

has swallowed its find will even be pursued by its fellows, and compelled to disgorge.[1] We have, indeed, here among the Gulls themselves the beginnings of the parasitic habit so highly developed in the closely related Skuas.

Live fish forms a comparatively small part of the food of Gulls, except when they happen to be following shoals. Dead fish they get occasionally either stranded on the beach, or thrown away by fishermen, and may be seen eating the same in company with other species, notably of the Crow Family, a temporary association by no means favourable to friendly relations. The Crows appear generally to get the worst of the encounters that take place. Saxby once saw a herring-gull drive off a raven, which had ventured to approach and pick up one of some sillacks thrown from a cottage door, and not only was the intruder compelled to drop its booty, but it was chased for nearly a mile, all the time crying out "as though it were being murdered."[2]

A large part of the food found by Gulls on the sea, especially in harbours and estuaries, consists of floating matter of all kinds, a biscuit thrown over the stern of a steamer, a lump of grease, or a dead dog. They carry their scavenging operations inland, and may often be seen in hundreds on and over the refuse-heaps near large towns. I was able, at some miles distance, to locate through a field-glass the position of the refuse-heaps near North Berwick by the living canopy of gulls formed above them when the birds were disturbed by the workmen in charge. Here again the Gulls come into conflict with the Crows, and, in the attempts made by one to snatch food from the other, it has been observed that the former have the advantage owing to their ability to swallow at one gulp a morsel that a rook or hooded-crow finds it necessary to divide, the delay giving the nearest gull ample time to make a snatch.[3]

Inland visits for food other than refuse are frequent, especially in

[1] E. Selous, *Bird Watching*, p. 118.   [2] *Birds of Shetland*, p. 340.
[3] *Journal für Ornithologie*, 1900, 347 (Wüstnei).

Plate 105

Lesser blackbacked-gulls and herring-
gulls (right) and Iceland-gull (left)

By A. W. Seaby

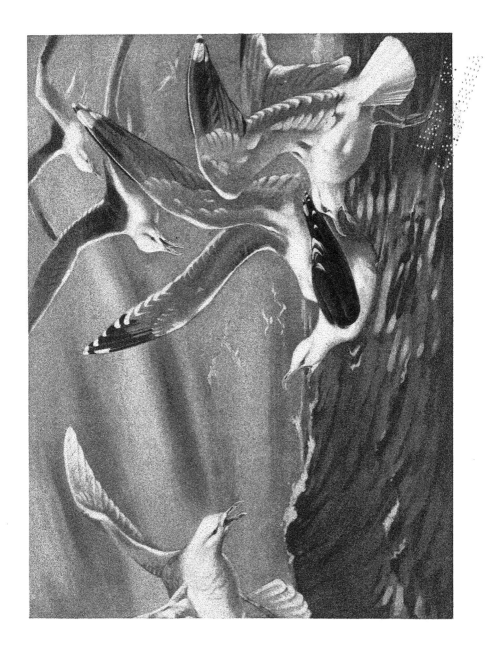

hard weather. Both the common and herring-gulls may often be seen following the plough, sometimes by themselves, sometimes with other species, such as rooks, starlings, and blackheaded-gulls. I have not seen the blackbacked species following the plough, and can find no record that they do, but they are not infrequently to be seen on the ploughed fields hunting for worms and insects like their congeners. They are, all four, great hunters of small rodents—rats, mice, voles, moles, and young rabbits. They hunt for these on field and moor, as well as for less legitimate game. Occasionally they pick up grain that is left uncovered, and make raids upon the turnips. Saxby saw a field of turnips half destroyed by herring-gulls. The birds eat the inside of the roots, leaving the outside untouched. As each bird made its way into the body of a turnip, its head naturally disappeared from sight, but not for long; it fed much like a tit at a cocoa-nut, dive, peck, a hasty withdrawal, nervous look round, another peck, and so on. The birds seemed fully aware of the temerity of their proceedings.[1]

The favourite feeding-grounds of the larger Gulls are the flats laid bare by the receding tide. There, in ooze beds or rocky pools, they hunt for starfish, crabs, whelks, mussels, and the like, and stranded carrion. Both the herring and the common species have been seen dancing on the mud or sand, like the blackheaded-gull, in order to stir up whatever eatables it might contain, including worms. Of a herring-gull kept by him in captivity Montagu writes: " When the weather is mild and the ground moist, it is amusing to observe its method of catching worms by a perpetual trampling on the same spot, turning about in all directions, and eagerly examining for those that rise out of the ground, which are instantly seized, and the same work is renewed. Similar means are frequently used by fishermen to procure worms for bait; but it would hardly be conceived that the slight pressure or concussion occasioned by the trampling of so small a body as a gull should force the worms from their retreat, but such is

[1] *Birds of Shetland*, p. 340.

the fact."[1] The two species mentioned also fly into the air with crabs
or shell-fish and let them drop from a height in order to break them.
A dozen herring-gulls or more have been observed stationed at regular
intervals of about a hundred yards from one another, all busily "shell-
dropping."[2] A common-gull has been noted to drop the same shell
ten times before breaking it.[3] Whether the lesser blackback drops
shells or practises dancing for its food is not recorded. The greater
blackback lets drop from a height the puffins it seizes, but what
purpose is served is not clear, for the unhappy victim is previously
banged into an almost lifeless condition and partly eviscerated.[4] This
gull has not been recorded as being in the habit of dropping shells
or crabs; it certainly has no need to drop the latter, for its usual
method, noted by Mr. A. H. Patterson, at Breydon, Norfolk, is to kill
the crab with a blow of its bill, then swallow it whole. An individual
watched by Dr. Heatherley used to carry its crab to a pool and
wash it before swallowing.

During the breeding season the three larger species, and to a
lesser degree the common-gull, devour a large number of the eggs
and young of other species nesting in their neighbourhood, such as
puffins, guillemots, shearwaters, kittiwakes, blackheaded-gulls, ducks,
and game-birds. Both the lesser blackbacked and the herring-gull
are in the habit of thrusting their heads down the puffin burrows to
draw out the chicks, which they peck into a condition of convenient
inactivity and then swallow whole.[5] The lesser blackbacked-gulls are
not content to devour the chicks, but also make short work of the
parents when they can catch them, which they succeed in doing by
the simple device of watching at the mouth of the nesting-hole.
When the unsuspecting puffin issues forth, it is seized by the neck,
shaken and banged about, and finally eviscerated, the opened body

[1] *Dictionary of Birds* (Newman's edit.). Cf. Macgillivray, *History of Birds*, v. 578 (common-
gull), 547 (herring-gull).
[2] Patten, *Aquatic Birds*, p. 427. See also Naumann, *Vögel Mitteleuropas*, xi. 248 (Jourdain,
etc.).   [3] Coward, *Fauna of Cheshire*, i. 430.
[4] C. J. King, *in litt.*   [5] F. Heatherley, *in litt.*

being left lying about.   The shearwaters are caught in the same way. Mr. C. J. King, to whom I am indebted for this information, and who has often watched the lesser blackbacks and their congeners engaged in their murderous work, collected, one year, on a patch of ground about 50 to 60 feet across, a heap of about thirty of these victims and photographed them.   This photograph, which I have, makes it clear that the gull contents itself with the internal parts, which it draws through a hole in the abdomen, leaving the rest of the body intact.   No doubt if puffins and shearwaters were not so abundant, the lesser blackback would not be so fastidious.   The great blackback shows itself at times equally fastidious, but, if indisposed to play the gourmet, it has this advantage over its congeners that it can save itself much trouble by swallowing its bird whole, a fact sufficiently proved by the presence of the heads of puffins and shearwaters among the castings of indigestible matter, fur, shells, bones, etc., which, like its congeners, it is in the habit of disgorging.   A good photograph of these ejected heads will be found on Plate xlvi. (p. 130).[1]   The ability of the great blackback to swallow comparatively large birds has, moreover, been proved by post-mortem examinations.   One was found to have swallowed a redshank whole, legs and all, another a little-auk, which was so slightly damaged by its incarceration that it was preserved by a taxidermist.[2]   The ventral capacity of the species may be estimated from the fact that six full-sized herrings were disgorged by one and the same bird.[3]   The great blackback's method of capturing its victim is more open and direct than that of the herring-gull and lesser blackback.   It suddenly pounces upon a puffin that happens to be off its guard, seizes it by the neck, shakes it as a dog does a rat, and lets it go.   The puffin, if not sufficiently injured, may profit by its release to fly off, and it sometimes escapes.   As a rule, the unhappy creature is seized again and again, and shaken till it can scarcely move.   The gull finally picks it up, flies up a hundred feet or so, and lets it drop

[1] The heads were found in castings examined by Mr. C. J. King in the Scillies.
[2] Nelson, *Birds of Yorkshire*, p. 684 ; *Nature*, 1895, 121.   [3] Nelson, *loc. cit.*

on the rocks. It then descends in a leisurely manner—for it has no fear that any other gull will venture to dispute its rights—and, if not very hungry, merely eviscerates its victim and devours the internal parts.[1] The great blackback occasionally spoils the spoilers. Mr. C. J. King informs me he has seen a pair of these birds tear in two the chick of a lesser blackback, and each swallow its half.

The herring-gull has not been observed to eat adult puffins and shearwaters, but it is difficult to believe that it is less enterprising in this respect than its fellows, the more so as it is known to seize fully fledged young kittiwakes, simply throttling them, and then tearing open the soft, warm breasts.[2]

It is perhaps hardly necessary to add that wounded birds, ducks, game-birds, and the like receive little mercy if they chance to be detected. The same applies to the small migrants that reach our shores on weary wings. They are promptly seized and swallowed whole.

After the recital of these gruesome gastronomic feats, it comes almost as a relief to know that the gulls themselves are victimised by Skuas. The latter chase and force them to disgorge their last meal, which they then snap up and swallow in mid-air. From these attacks the great blackbacked-gull seems to be exempt: it has apparently only the eagles to fear,[3] and human beings. The latter consume every summer bucketsful of gull's eggs, and they are said to be good eating. One lighthouse-keeper told me he was equal to a breakfast of three to four of those of the lesser blackbacked. Personally I never got beyond the first spoonful, the flavour being too strong for my taste.

The eggs are not always taken for such legitimate purposes, as the following incident narrated by Mr. J. M. Boraston will show.[4] He happened to be on Puffin Island one day when it was visited by a Field Naturalists' Club. In reference to some of the members of this

---

[1] C. J. King, *in litt.*   [2] E. Selous, *The Bird Watcher in the Shetlands*, pp. 308, 314.
[3] Naumann, *Vögel Mitteleuropas*, xi. 265.   [4] *Birds by Land and Sea*, p. 244.

Plate 106

Great blackbacked-gulls

By A. W. Seaby

Club, which it is only fair to say must not be taken as typical, he writes: "I had never before seen grown men invade a bird-haunt like savages, rushing from nest to nest with excited shouts—beside themselves, in fact, with the wealth of plunder lying at their feet. One fellow, more methodical in his barbarity, went about testing the state of the eggs by working a knife-blade through the shells: if the egg contained a chick in an advanced state of development, the blood which appeared showed that it was useless for blowing purposes, and it was returned to the nest. Upon my expostulating at this barbarity, he expressed regret that he 'had no stamp-paper!' If the egg gave signs of having been freshly laid, it was stored with a growing collection in a wicker-basket he carried. . . . The eggs so gathered were brought down to the beach for final test; if they floated, they were too far advanced for cabinet purposes, and were stoned down." [1]

Such barbarity is far more revolting than the disembowelling of puffins above described, for it serves only an idle and childish purpose; the gulls, at least, confine themselves to killing for food, in which they show themselves vastly more civilised than mankind.

In August or earlier, they quit for the most part their breeding-haunts, and are found during the autumn and winter in various parts of the British Isles, both inland and on or near the coast, or out at sea. They are usually seen in flocks or small parties, either separate or associated with other species, the least gregarious being the great blackbacked. But flocks of thirty or more of these are at times observed. The relations between young and old at this season have yet to be closely studied. They are not always very cordial, and this is probably due in the first place to the determination of the parents to make the young understand that they must shift for

[1] Mr. Jourdain tells me that a small incision with a pin made in an egg will not prevent it hatching, and that if covered with stamp-plaster no damage will ensue. This fact is not given in order to vindicate the behaviour of the gentleman referred to in the text; he used a knife-blade, and, on his own showing, "had no stamp-paper."

themselves instead of continuing to beg for their food. I used to
watch a small company of herring-gulls that took up their winter
quarters in St. James's Park.   Near them were four or five immature
birds of the same species, who would gladly have joined their elders
had they been permitted to do so.   But no sooner did one of them
make advances than an old bird would begin to edge towards him with
an air of meditative detachment from mundane affairs that might well
have deceived the proposed victim had he not already been taught its
meaning by experience.   But such is the fascination exercised by
exclusiveness, that the young gull would persist in staying up to the
moment that the approaching bill was suddenly aimed at him like a
dagger, and he persisted in returning.   The only time he actively
resented this treatment was at meal-time; then his protests were
wide-beaked and his conduct irreverent.

There is evidence to show that, as autumn draws on to winter, all
four species make fairly general but irregular southward movements,
which may or may not extend beyond our shores.   For instance, a
young herring-gull ringed in the summer in Aberdeenshire, was recap-
tured on January 30 near Manchester.   At an earlier date, October
3, another was caught farther south, at Hunstanton, Norfolk, while a
third was still at Aberdeen on November 15.   About the same period
a young lesser blackbacked-gull, ringed at the Farnes in June, was
taken as far south as Portugal.   The few exact facts that we possess
about these movements are due to the modern practice of placing
rings on the legs of birds at the breeding-place; they suffice to
assure us that exceedingly interesting results may be expected in
the future.

GLAUCOUS-GULL

[F. C. R. JOURDAIN]

As a British bird the glaucous-gull is only known as a winter visitor, principally to our eastern coasts, and much commoner in the northern part of Great Britain than in the southern half. It is an annual visitor to the Shetlands, which seem to lie on its southward line of migration, but is comparatively rare on the Orkneys. Saxby tells us that small flocks, composed of young and old birds, but chiefly the former, appear about the middle of October. Some of the young birds may be seen throughout the winter months, but the old birds disappear shortly after their arrival.[1] These flocks travel down the east coast, but seem to have a definite route, not following the indentations of the coast, but working from one outlying point to another. On account of this habit most of our records come from certain localities on their flight-lines. It must not be supposed that the numbers which visit us are at all uniform, for they vary extraordinarily from year to year. Edmonston observed a flock of over a hundred birds in November 1820 in Balta Sound in the Shetland Isles ; and in November 1864, in the face of a gale of wind, Saxby saw a flock of about a hundred and forty birds flying over. In 1871-72 immense numbers of this species, and probably also some Iceland gulls, visited the east coast of Scotland, and were especially noticed in the Firth of Forth and also at the Tay mouth, the Eden estuary, and the coast of north-east Fife.[2] In the winter of 1872-73 a still more extensive migration took place. Mr. Harvie-Brown describes how, at St. Andrews, he saw "literally hundreds—principally glaucous-gulls— that winter, streaming along the tops and under the shelter of the sandhills which fringe the golf-links, flying low and almost touching

[1] Saxby, *Birds of Shetland*, p. 349.
[2] J. A. Harvie-Brown, A *Fauna of the Tay Basin and Strathmore*, p. 338.

the tops, and hugging the shelter these afforded from the gale from seaward that was blowing." It may be a coincidence, but this winter was remarkable for the enormous shoals of sprats (*Clupea sprattus*) which swarmed in the narrow tideway of the Upper Forth at Kincardine, as well as higher up, to such an extent that they were shovelled out on to the adjoining fields as manure in hundreds of tons. It was noticed that the great invasion of Pomatorhine and other skuas in October 1879 was also accompanied by the presence of vast shoals of herrings and sprats off our east coast,[1] so that it seems probable that the two events were connected with one another in each case. Other important visitations of this species took place in 1876-77 and 1894-95, and during the terrible storms of January 1881 some forty-five or fifty specimens were brought into Yarmouth by the fishing boats, probably for the most part from the Dogger Bank.

By far the greater number of these birds are immature ; it is only rarely that a specimen in adult plumage is recorded, but the great preponderance of immature birds is partly accounted for by the long period which this species takes to assume the fully adult plumage. There is no reasonable doubt that it does not breed during the first four years of its life, but the summer home of these immature birds is practically unknown. Dr. Alfred Walter, while stationed at Whales Point Harbour in East Spitzbergen from May 29 to June 7, 1889, noticed that on his arrival all the glaucous-gulls were old birds in full plumage. On May 30 the first immature birds appeared, a flock of six and then a single bird, all heading due north. From this date onwards many young birds in all stages of plumage, sometimes accompanied by old birds, passed over in flocks, the largest of which numbered from twenty to thirty individuals.[2]

Even in the immature state the glaucous-gull is not difficult to recognise. In size it equals the great blackback, and has a steady, soaring flight. The wings are less bent at the carpal joint than in the other gulls, giving them the appearance of being more extended.

[1] Stevenson, *Birds of Norfolk*, iii. 351.   [2] *Journal für Ornithologie*, 1890, p. 243.

Plate 107

Glaucous-gulls

By Winifred Austen

Naumann compares the flight to that of the common and rough-legged buzzard, while Gray describes it as "soft, sedate, and owl-like." When seen at close quarters, the conspicuous white or whitish quills form a useful distinguishing character, especially as it is not particularly wary, and will allow the observer to approach closely, as a rule. The only other white-winged species is the Iceland-gull, which is, however, a smaller bird. In full breeding plumage it is practically all white except the pale grey mantle and wings, but, as Saxby points out, at a distance it appears to be a dull, creamy white all over. He also notes that in the water it sits more buoyantly than the herring-gull.

When on the wing it generally flies high, and on migration is almost always seen in company. It is less curious and inquisitive than the other gulls, and does not show the same wish to closely inspect every visitor. In the Shetlands, where it is little disturbed, it will allow a very close approach. Saxby says that while wandering along the shore he has more than once come across an old bird feeding amongst the weed, and has seen it merely walk into the water and swim about a little until he was past, and then wade leisurely ashore and renew its search for food.[1] With the possible exception of the kittiwake, none of the gulls which commonly visit us can be said to be pleasant or desirable companions to any living creature smaller or weaker than themselves; but though accounts of this species vary, there seems little doubt that its disposition is less savage and bloodthirsty than the great or lesser blackback or the herring-gull. Its bearing is sedate and lacking in vivacity, and it does not warn other species of the presence of danger, but contents itself with quickly flying away. When pressed by hunger, of course it is greedy and voracious to a degree, but, where carrion is obtainable, it does not seem to pay much heed to smaller birds, except when wounded or injured in some way, though there is little doubt that the colonies which breed on the top of the great loomeries in Novaya

[1] *Birds of Shetland*, p. 350.

Zemlya and other places subsist chiefly, if not entirely, on their helpless neighbours' eggs and young. Scoresby says that the fulmar, the ivory-gull, and the kittiwake all retire when the burgomaster descends on its prey. W. H. Hudson noticed that a glaucous-gull when approached by another gull while feeding, "looked fixedly at him a couple of moments, then drawing in its head, suddenly tipped its beak upward—an expressive gesture, corresponding to the snarl of a dog when he is feeding."[1] The significant movement produced a marked effect on the other gulls. On the other hand, a nest of the glaucous-gull has been observed surrounded at a respectful distance by eider-ducks sitting quietly on their eggs,[2] and Kolthoff has seen an eider-duck drive off a pair of these gulls. The Rev. A. E. Eaton also says that the glaucous-gull stands in some awe of the fulmar in the water, while on the ice the ivory-gull is treated with respect. Dr. Saxby even goes so far as to describe it as good natured—rather a strong expression to use of one of the larger gulls,—and adds that it is the least meddlesome of its tribe, though a match for the great black-back itself, and far more than a match for any of the rest.

On an average the glaucous-gulls have left the Shetlands on their way to their breeding-haunts by the middle or end of March, but they are somewhat irregular and uncertain in their movements. Though a single bird is said to have been seen at Bear Island in February, Mr. A. Pike did not observe it on Spitzbergen till March 26, while W. H. Neale records the first from Franz-Josef Land on April 22, and Bunge saw the first at Horn Sound on April 4, and did not find it common till May 5.

They resort to the same breeding-places year after year, and probably also make use of the remains of the former years' nests. The tendency of this species is to breed in colonies, but the nests are not placed close together, and isolated pairs are frequently found nesting considerable distances away from any others. The nesting-sites vary according to the locality; thus on Spitzbergen many nests

---

[1] *Land's End*, p. 23.   [2] *Zoologist*, 1874, 3811.

are placed among the shingle and drift on the beach above high-water mark, or so low down on the cliffs that they can easily be reached, while others are placed on low rocks and islets, and one nest was found by Eaton among the upturned roots of a dead spruce fir.[1] In the great loomeries of Novaya Zemlya Pearson found the glaucous-gulls occupying the topmost ledges of the great ranges of cliff, from whence they could look down on the myriads of breeding guillemots below them.[2] On the other hand, on Korga Island, off the Murman coast, he found a large colony breeding on the sand-dunes, which reached to a height of some thirty or forty feet,[3] and Seebohm also found young birds on the sandy, low-lying Golaievski Islands, near the mouth of the Petchora River. Where there is plenty of material available, the glaucous-gull builds a big substantial nest of some two feet in diameter and 3½ to 5½ inches high, composed of tang and other seaweed, moss, and dead grass pulled up by the roots. Kolthoff noticed blossoms of *Saxifraga oppositiflora* in the nest, and Le Roi found *Cochlearia officinalis* similarly used on Bear Island. On sandy islets where there is little in the way of material, the nests are mere hollowed heaps of sand lined with a few tufts of seaweed, while on Prince Charles Foreland Kolthoff found nests composed entirely of stones about the size of a pigeon's egg.

In most cases fresh or slightly incubated eggs may be taken in the first or second week of June, but Römer and Schaudinn found many young in down in Spitzbergen on June 13, so that in some cases the eggs must be laid by mid-May. Off the Murman coast Pearson found the eggs fresh or only slightly incubated on June 27. The eggs are from two to three in number as a rule, and Pearson found that all three young were frequently hatched out, and that the third egg was not necessarily unfertile, as is often the case in Spitzbergen. Four eggs have occasionally been found in one nest, and in Spitzbergen this seems to have occurred more than once, for both Römer

[1] *Zoologist*, 1874, 3811.
[2] H. J. Pearson, *Beyond Petsora Eastward*, p. 163.
[3] *Three Summers in Russian Lapland*, p. 105.

and Schaudinn and Koenig record clutches of four eggs. Although in some cases indistinguishable from those of the great blackbacked-gull, they are frequently a much blunter and more rounded oval in shape. Like most gulls' eggs, they are found in varying shades of olive-brown or bluish green, and are rather sparingly blotched, streaked, and spotted with dark maroon or blackish brown and violet-grey shell-marks. On Koenig's expedition a light bluish green clutch, only slightly marked with a few shell-marks and greyish brown blotches, was found on Bear Island. Formerly it was believed that the red eggs of the herring-gull (and perhaps also of the great blackbacked-gull), which have occasionally been taken on the islands off Vardö in Norway, belonged to this species, but it is now generally agreed that it does not breed in Norway at all.

Of the habits of this gull during incubation and while rearing the young we know practically nothing. Yet it has bred more than once in the Zoological Gardens at London, and two young birds which were hatched there on June 24, 1868, were still alive and well in 1879. There is, however, no notice of this interesting event in the *Proceedings of the Zoological Society* for 1868, and *Larus glaucus* is not even mentioned in a list of birds which have bred in the Gardens, which appeared in the *Proceedings* for 1869. These birds, however, enabled Howard Saunders to watch their changes of plumage to maturity, and to satisfy himself that the almost uniform creamy or even perfectly white plumage is assumed for a very short time just before the autumnal moult at which the pearl-grey mantle is assumed. It was a bird in this stage which was described as a new species by Richardson, under the name of *Larus hutchinsi*.

Mr. F. J. Jackson tells us that on his approaching a nest, one of the birds scattered some of the nest material over the eggs, no doubt with intent to conceal them. They showed considerable courage in defending the nest, swooping down within a foot or two of his head, and uttering loud screeches as they passed.[1] Fischer says that both

[1] F. J. Jackson, A *Thousand Days in the Arctic*, p. 391.

sexes will pursue the nest plunderer for hours.[1] As brooding-spots are found in both sexes according to Naumann, it is probable that they share the duties of incubation. He also gives the period as four weeks, which is probably approximately correct.

The downy young are, according to Seebohm, like those of *Larus affinis*, but the dark markings on the back are fewer and fainter. Saunders describes them as stone-grey, with a slight tinge of yellowish buff above : the head spotted with black, and the back mottled with ash-brown. In the Spitzbergen group most of them are hatched by the end of June or early in July. They leave the nest as soon as they can fly, and walk freely about the rocks. A young bird taken from the nest on Jan Mayen was visited for several days by its parents, and on one occasion followed them to the sea for six hundred paces. This bird usually became excited on hearing the call of its parents, while two others, their heads bent sideways, listened with indifference.

Hardly anything comes amiss to this powerful bird in the way of food. When a whale is being "flensed" he is in close attendance, swooping down on the floating fragments and carrying them off on the wing. Colonel Feilden relates how two reindeer shot by him were reduced to hides and skeletons in an extraordinarily short space of time by these birds. Almost any kind of carrion, and the excrement of the walrus, seal, and polar bear, is also devoured. When the carcass of one of the larger mammals is discovered, the eyes are first picked out and access thus obtained to the brain. This method of procedure is shared with other large carrion-eating birds, such as the Ravens and Vultures. In the breeding season vast numbers of eggs and young birds are taken, and it has been seen in hot pursuit of a little-auk, a young Brünnich's guillemot, and a wounded ivory-gull. One killed in Norfolk was eating a dead coot, and another contained a whole golden-plover, while a bird shot by Captain Ross disgorged a little-auk, and on being opened another quite whole was found in its stomach ; but healthy birds are rarely interfered with. Fish of all

[1] *Zoologist*, 1890, 49.

kind, alive or dead, are freely eaten, and the lump fish, *Cyclopterus lumpus*, is especially sought for. Among Crustacea we may mention *Cancer pulex* and *C. araneus*, and among Mollusca *Venus islandica* and *Pecten islandicus* as forming part of its diet, as well as a large species of Decapod (*Hyas*) and Holothurians. Hagerup states that in spring, when the flocks of eiders are diving for mussels off the Greenland coast, the glaucous-gulls may be observed swimming or flying among them, and as soon as a duck comes to the surface with a mussel, a gull attempts to seize it (*Auk*, 1889, p. 214). But this is not all, for even when animal food is plentiful it likes to vary it with seaweed. Saxby has taken as many as five species of Algæ from the stomach of one bird, chiefly *Alaria esculenta*, while Naumann says that it is said to eat the berries of *Empetrum nigrum*. With such varied tastes it is perhaps not surprising that Saxby found only one of all that he examined to be in poor condition; usually they were so fat as to cause much trouble in skinning, and the fat always smelt strongly of whale-oil. The more indigestible portions of the food are thrown up again in the form of pellets, and Schalow gives some interesting details of the contents of some of these obtained in Spitzbergen.[1] One huge pellet, nearly 8 inches long and 1 inch wide, contained the remains of a young ivory-gull, which had been swallowed whole. Another was globular-shaped, and measured about $2\frac{3}{8}$ inches long and $1\frac{1}{8}$ inch wide. As Schalow remarks, it seems inconceivable how a pellet over 6 inches long and $1\frac{3}{4}$ inch wide can pass through the gullet without injury to the bird.

The notes of this species have a great resemblance to those of the other large gulls, such as the great blackback. Le Roi writes the call-note as a loud "*kau kau kawkawkaw*," which is probably the same sound which von Heuglin syllables as "*gogäu, gogäu*," and Naumann as "*güowüüü*." When the breeding-place is invaded a shrill "*gagagak*" or "*gogogok*" is incessantly uttered, probably the former being a remonstrance by the hen and the latter by the cock bird. Other

---

[1] *Journal für Ornithologie*, 1899, p. 378.

notes which are uttered on the wing are "*kuija, kuija, kija,*" or a short "*kaea kaea.*"[1]

Towards the end of October the breeding-grounds in Spitzbergen are forsaken. Arnold Pike noted flocks for the last time on Amsterdam Island on October 13, 1888,[2] and Bunge saw one or two as late as October 26 and 27; but by the beginning of November the last burgomaster has left the ice-bound coast behind him, and winged his way southward to the open sea and sheltered fjords of Iceland and the Færoes, or the voes and firths of the Shetlands. The Austro-Hungarian expedition to Jan Mayen, however, had a somewhat different experience, and both old and young birds were seen as long as the sea was free from ice, *i.e.* till December, while a young bird was shot at sea on January 22, 1883, and again on March 17.[3]

ICELAND-GULL

[F. C. R. Jourdain]

The material available for the life-history of this gull is at present far from complete, chiefly because most of those who have had opportunities of observing it at home have contented themselves with remarking that in its habits it resembles the glaucous-gull. Very probably this is the case, but in studying one cannot afford to take anything for granted, and it may be that closer investigation will reveal essential points of divergence, which in the course of time have resulted in the permanent differentiation of the two species.

In size the Iceland-gull is decidedly a smaller bird than the glaucous, but the wings are much longer in proportion. Saxby says that on the wing it can be recognised at any distance by its long and pointed white wings, and by a peculiar roundness of body. The flight, too, as might be expected, is more airy and buoyant, and not so owl-

[1] Le Roi, A*vifauna Spitzbergensis*, p. 192. [2] Chapman, *Wild Norway*, p. 344.
[3] *Zoologist*, 1890, 49.

like as that of the glaucous-gull, and Harvie-Brown says that when flying low or against a dark cloud, the white primaries show like a narrow strip of silver along the wing. When at rest, the same observer states that the Icelander can be distinguished by its neater and more slender appearance, standing higher on its feet, and carrying the tips of its long wings at the same level as the end of the tail, while the glaucous-gulls carry the tips of the wings higher, and in consequence have a less trim and tidy appearance. It seems also to be more active and energetic than its larger neighbour, in this respect resembling the herring-gull; and Faber says that it is not afraid to fight with equal or even superior antagonists for food.

As far as the British Isles are concerned, it is a fairly frequent, but very irregular, winter visitor to the north. In the Shetlands and Orkneys it occurs annually, but only in small numbers, and occasionally ranges down the east coast of Scotland in some numbers. During the great invasion of glaucous-gulls which took place in the winters of 1871-72 and 1872-73, which has already been described, considerable numbers of Iceland-gulls were also seen. On January 4, 1873, six adult birds were identified by Messrs. R. Gray and J. A. Harvie-Brown in the Firth of Forth. After that date they occurred more frequently, and on some days were even more numerous than the glaucous-gulls. On the 13th, within a few seconds, Harvie-Brown counted no fewer than twelve adult birds, flying low against the wind, and distinctly showing the white primaries. From this time on they were constantly in sight, two, three, or more at a time, all flying away inland and alighting on a ploughed field on Dunmore estate, till towards the afternoon not a single Iceland-gull was visible over the Firth. Comparatively few birds out of this body were noticed on the Tay. Along the east coast of England it is always rare, and the few that are met with are generally immature. Considerable numbers, however, reached the Cornish coasts in the early part of the year 1873, and, after a succession of stormy days, both adult and immature birds were met with on the South Devon coast in the winter of 1874-75.

These birds may have reached us by the west coast of Great Britain, as some usually migrate along the west coast of Scotland; and Dr. Salter says that in some winters it is not rare along the coast of Cardigan Bay. Thus, in January and February 1892, it was observed in some numbers on the north and west coasts of Scotland, and this migration extended along the shores of North Ireland, where it had not previously been noticed in any numbers, though since that date it seems to have occurred on many occasions.

On the way north an occasional straggler may be met with late in the spring, but though recorded from the Humber as late as April 18, the great bulk of our winter guests leave the Shetlands towards the end of March. In Iceland they stay later, and Faber says that though their numbers decreased towards the end of April, they had not all disappeared till the end of May. To some extent their movements here are regulated by the ice, for in the spring of 1821, when the fjords of the northern coast were choked by drift ice from Greenland, they stayed on the south coast till the middle of May, and then left it entirely, to proceed to their breeding-haunts. When we come to trace out the breeding-range of this species, at first sight it appears to be wholly a nearctic breeding species. It is true that it ranges in winter to Norway and Finland, and von Pelzeln identified a specimen from Novaya Zemlya, while Middendorff thought that he recognised it on the Taimyr, but Saunders ascribed all the Bering Sea and North Pacific records to the glaucous-gull. On the other hand, we know that there is a strong colony on Jan Mayen, and that it breeds in great numbers in Greenland, while the principal localities in Arctic America where it has been found nesting are given in the "Classified Notes." But outside this area, which undoubtedly represents the chief breeding-grounds of this species, there is some evidence that it occurs sporadically in small numbers over a much wider area. Thus E. W. Nelson states that it ranges not only across Bering Straits, but along the east coast of Arctic Siberia to Ice Cape and Wrangel Island, and Bunge met with it on the Lena in latitude 71° N. Moreover, M. N. Ssmirnow, during his voyage in the ex-trawler *Pomor* from March to August 1901, kept a diary of the birds observed by him while cruising in the Kara Sea and the Arctic Ocean between Novaya Zemlya and the North Cape (sometimes known as Barents Sea). Off the Murman coast and the entrance to the White Sea, Iceland-gulls were observed at various dates between March 20 and April 17. Still more remarkable is the occurrence of this species off Kolguev on June 18, 23, and 25, while on the southern island of Novaya Zemlya not only were birds observed early in July, but a nest, which apparently contained eggs, on an inaccessible pinnacle of slate, was found on July 3, and both parents shot.[1] As glaucous-gulls were also met with at most of these localities, it is difficult to understand how there could possibly be any confusion of the two species in this case. On the other hand Colonel Feilden, who accompanied Mr. H. J. Pearson on his voyages to these seas, and who is thoroughly acquainted with the Icelander, informs me that neither he nor his companions ever saw a single gull which could even doubtfully be ascribed to this species. At any rate, the question of its occurrence in the Palæarctic region is one which deserves investigation, and, while further corroboration is needed before M. Ssmirnow's identification can be accepted, the problems of distribution cannot be solved by merely ignoring them, as is too frequently the case.

With regard to the breeding-habits of this gull, little has been recorded. On Jan Mayen the nests were placed on low ledges, often scarcely projecting above high-water mark,[2] but there is evidence that it is also frequently found breeding on the tops of lofty cliffs. Thus in Greenland about a thousand pairs breed on a cliff overlooking the fjord near Ivigtut, about 2500 feet above the water, according to Hagerup. If the nests ascribed by Dall to this species are correctly assigned, the nests on the Lower Yukon were mere depressions in the sand. The eggs are two, or more frequently three, in number, and are

---

[1] *Ornithologische Jahrbuch*, 1901, p. 201.   [2] *Zoologist*, 1890, 50.

much smaller than those of the glaucous-gull. The average of 46 eggs is 2·69 × 1·89 in. [68·4 × 48·1 mm.], and in ground-colour they vary from light olive and grey-buff or greenish, to olive-brown, blotched and spotted somewhat evenly with dark umber-brown markings, and grey-violet underlying blotches. In some eggs the blotches and spots tend to form a wreath. All detailed information as to the share of the sexes in nest-building and incubation, as well as the length of the incubation period, is still lacking. The eggs, in Greenland, are generally laid between May 25 and June 28.

The young of this species pass through similar stages of plumage to those of the glaucous-gull, but the pure white stage, which has been rarely met with in the case of the glaucous-gull, on account of its brief duration, was unknown in this bird until described by Eagle Clarke in the *Proceedings of the Royal Physical Society of Edinburgh*, 1899, p. 164, and, as in that species, is only retained for a short time, just before maturity is attained in the fourth year.

Like the other large gulls, it is practically omnivorous, and in the Arctic regions subsists largely on carrion, the excrement of large mammals, and wounded or injured birds. It also devours fish, molluscs, and crustaceans, and Saxby notes that it is more partial to vegetable food than some of its congeners, often resorting to the fields, where it is seen in attendance on the tethered pigs, possibly in search of worms. The stomachs of birds examined by him contained oats, vegetable fibre, and quartz. Faber states that in Iceland they come to the house doors in winter to pick up the entrails thrown away by the inhabitants, and fight fiercely for them with the ravens. To the seal-hunters they are useful as indicating where seals may be found, by continually pursuing them overhead as they swim below, hovering over them while in pursuit of fish, and swooping down upon their prey when driven to the surface. In the same way they will follow the track of the cod-fish, to feed upon the fish and other organisms pursued by them. On the whole, it may be described as more of a fish-eating gull than a flesh-feeder.

The notes are said to differ considerably from those of the glaucous-gull. Hantzsch writes the cry as "*gi, gi, gi, grrr*," and says that it has notes like those of *L. glaucus*, but softer, "*Gag gagag gogogogog gigigigig.*" Faber writes them more briefly as "*ik-knirrrr, giow*" (like that of *L. marinus*), and "*hoo*"; while Saxby, without venturing to express the note by letters, says it has a character of its own, and proceeds to compare it with that of the goose.

Soon after the middle of September the first arrivals from the north appear in Iceland, and from that date onward some may be seen throughout the winter till their departure about the end of April for their breeding-grounds.

THE KITTIWAKE.

[F. B. KIRKMAN]

This, the most fascinating of our Gulls, differs in marked particulars from the other members of its subfamily, and is placed in a genus (*Rissa*) apart. One of its distinctive peculiarities, the vestigial first or hind toe, is indicated by its specific name *tridactyla* (Greek: *tri*, thrice; *daktulos*, a digit); all that usually remains of this toe is a warty excrescence with a small claw. In all other genera of Gulls it is developed though short, too short indeed to be of much, if of any, use. The gap that divides *Rissa* from *Larus*, to which genus belong the species treated in the preceding sections of this chapter, becomes still more apparent when their young are compared. The down-clad nestlings of the latter have the head and back more or less conspicuously spotted, striped or barred with shades of black, brown, and buff. The nestling of the kittiwake has the head unspotted white with a grey tinge, and the back ash-grey, with *very faint traces of darker mottlings*, which may be regarded, like the hind toe, as vestigial. A glance at the photograph of the nestling kittiwake on Plate XLIV. (p.

116), and the nestling common and herring-gulls on Plate XLV. (p. 124), will make clearer the differences here noted. That the nestlings of *Larus* should retain a spotted and striped down is intelligible, seeing that for the most part they live on the ground, where their broken and irregular coloration has some protective value, in any case more protective value than a plumage which has no markings to break the outline of the body and so render it less conspicuous against its background. It is also intelligible that the nestling kittiwake can have no use for a similar coloration on the ledges of the sea cliffs or caves where it is invariably born, but why it should have developed its present coloration is at present inexplicable. The fledgling is also quite unlike those of *Larus* in certain respects, notably the dark semi-collar and the black on the wing-coverts. It is less unlike its parents than are they.

The breeding-range of the kittiwake lies chiefly within the Arctic Circle; its colonies are found all along the northern coasts and islands of America, Asia, and Europe. The largest, perhaps, is near the North Cape, on the cliffs at Svaerholtklubben, which numbers some millions,—so many that, when the birds are disturbed by a gunshot, they look, as they rush from the rock, like "a snowstorm in a whirlwind."[1] The most northerly known is in Franz-Josef Land, at about 80° N. latitude, where eggs were recorded by Nansen and Collett, but not till after June 17. On June 13 and afterwards single birds or small parties were observed daily as far north as 82° 20′, wherever there was open water, and by Sverdrup at 84° 52′. They were met with at 81° up to August 23, and a few degrees southward in flocks on the edge of the ice as late as September, and were the last birds seen in those regions.[2] They breed also in the Temperate Zone, including many localities in our Isles, and in winter range south to the Tropic of Cancer, and even beyond into equatorial waters.

At their British breeding-haunts they arrive usually in March,

[1] R. Collett, *Bird Life of Arctic Norway*; Seebohm, *British Birds*, iii. 343.
[2] Collett and Nansen, *Norwegian North Polar Expedition* (1893-95) *Scientific Results*, vol. i, chap. iv. p. 10.

but do not begin building their nests till May. Little has been, so far, recorded of their habits in the interval. It is, as might be expected, largely taken up by love-making, and with a certain amount of fighting arising from disputed claims to nest-sites. The kittiwake's manner of expressing its love has yet to be studied in satisfactory detail. I have personally only had opportunities of watching the species from early May, when it begins nest-building, till August. All I can find stated about its love-gestures previous to May is summed up under the general terms "billing" and "cooing," which lack precision, especially the latter. But as it is highly unlikely that the love-gestures of the birds in the early part of the season differ from those they make later, what is here said of the latter may be regarded as applying equally to the former.

Kittiwakes are very demonstrative in their affection. Mated birds may often be seen facing one another and uttering excitedly, as if their conjugal happiness depended upon it, the familiar *kitti-way-ĕks*, which give the species its name, and which may easily be imitated by pronouncing the word as would a parrot. They may accompany this outpouring with a variety of gesture. One pair, for instance, after bending down their heads, and, with beaks wide open and tongue conspicuous between the mandibles, kittiwaking on to the bare rock, levelled their heads and necks and kittiwaked at one another, and finally raised them and kittiwaked to the heavens. During the performance the neck was also moved from side to side, so that every point of the compass was favoured with the sound of their voices. This was on August 15, at the end of the breeding season. Again, they may often be seen rubbing beaks, in other words billing, and following the same by ecstatic kittiwaking, usually with the heads bent downward in the regurgitative manner common among other species of Gull. The wings are sometimes opened, sometimes closed, during these performances. The same outbursts of *kittiwakes* may follow the return of one of a pair to the ledge where its mate is keeping guard or incubating. One may often, indeed, tell when the return is about to take place by watching the bird on the ledges. Before its mate actually alights, while it is still some yards away, it will suddenly open wide its beak, and *kittiwake* a hearty welcome, at the same time displaying the brilliant orange hue of the inside of its mouth, brilliant, indeed, like the inside of a flame-coloured flower, delicate as fine porcelain, and made all the more gorgeous by its contrast with the yellow bill and the soft whites and greys of the plumage. The kittiwake, indeed, and many another bird have this advantage over us, that when they open their mouths to their mates there is at least something worth seeing if not worth hearing.

The most singular of the domestic amenities of kittiwakes—if one may for the moment class it as such—takes the form of a peculiar swallowing action that makes itself very visible externally, and gives the bird the appearance of trying to gulp down something that, if left to itself, would come out. A pair may often be seen for several seconds facing, but not touching, each other and "gulping" The action is sometimes accompanied by bowing and sometimes by kittiwaking. At Bempton, in May, it occurred frequently after the kittiwaking which accompanied the return of a bird to its mate on the nest ledge. On one occasion one seemed to disgorge, the other gulped, but neither on this nor on other occasions did I see anything in the nature of food or feeding. Like the other actions above described, except, of course, the billing, it is as often performed by birds when alone as when with their mates. It therefore represents a state of feeling which does not need the stimulus of another bird's presence. But what this state of feeling may be I do not know. The action itself originates, without doubt, in the actual swallowing movements made by both old and young, which will be described later.

Besides the *kittiwake* already mentioned, the species utters at least two notes, probably more. One is a clear, somewhat metallic sound, more or less like "*uck, uck, uck.*" I have heard it preceding the *kittiwake*, and it corresponds no doubt to that figured by Naumann as

"*dack*"! [1] It is also used by itself. The second is a petulant little infantile note, which I find figured in one of my note-books as *keee-kee-ke*,
in another as *kwee-kiouh*, neither of which, however, recalls to me even
faintly the actual sound. I have heard it following an outburst of
kittiwakes, and also interspersed among them. It was uttered by
birds when collecting nest material on the cliff, and by one was
intended to be a menace. It was also uttered by birds on the wing.
According to Naumann, it is heard only at the breeding-place. The
meaning to be attached to this and the other two notes has yet to be
precisely determined. The same, indeed, may be said of the notes of
all the Gulls.

Nest-building begins in the first half of May, or later according to
the latitude. At the Flamborough colonies the bird had in 1911 already
set to work by May 9. Any ledge on the face of a cliff large enough
to accommodate the nest will satisfy a kittiwake. If there are remains
of the old nest, the new is built upon it, and this again, if not all
washed away by wind and rain, may provide a foundation for the
next season's nest. In time, therefore, the structure may assume large
dimensions. As a rule, however, nests do not survive the winter,
though they are substantial structures. They are built mainly of
compressed mud and seaweed or grass, and lined with finer material,
usually grass, which is frequently added to during the course of the
season. The sides of the nest become thickly saturated with the bird's
excreta. One nest I saw at the Bass Rock in August had hanging
over the edge of it a dead kittiwake, so coated with white excreta
that it looked like a petrified bird.

Both sexes share in the building of the nest, and their method of
working, which I had ample opportunities of watching at Bempton, is
much more thorough than that of other Gulls. One bird usually remains
on the ledge while the other flies off for material, which it finds either
on the sea, the shore, or the earth-covered places on the face or top
of the cliff. At Bempton I saw, at a distance of a few yards, half
a dozen birds and more at a time on a wet clayey patch near the top

[1] *Vögel Mitteleuropas*, xi. 290.

of the cliff, into which they dug the clean yellow bills until loaded and coated with mud.   With this, or with tuft of grass tugged up by the roots, they flew off to their ledge, where they were welcomed with loud *kitti-way-êks* by their mates.   Each on alighting dropped its contribution, or, in the case of the more adhesive mud, shook it off and out of the beak on to the growing nest, but sometimes so energetically that bits were jerked over the edge of the ledge down into the sea.   The bird then trod the addition in with alternate movements of the feet, and also from time to time pressed it down with the breast.   The treading process lasted several seconds at a time, the little builder being evidently determined to make the structure cohesive and solid.   Its mate waited generally a few seconds to see the work well started, and then flew off for material, returning to be greeted in the same way. Sometimes the same bird went mud or grass hunting two or three times in succession.   Sometimes one would go off on a jaunt, or on a feeding excursion, or drop down to the sea either to bathe or to clean its muddy beak, which it did by rubbing it with its feet and splashing it about in the water.   Again, at times, a flock of birds together would suddenly and simultaneously cease work and fly out to sea a hundred yards or so, the outflight, like the sudden up-flights of the blackheaded-gulls (p. 144), being silent, while the return was all noise and excitement—a winged storm of *kitti-way-êks !*

The kittiwake frequently breeds on the same cliff with other species, usually guillemots and razorbills, also with gannets.   The relative nesting-position of these species is no doubt primarily deter-mined on each cliff face by the character of the rocky surface ; the broad open ledges, wherever they are, are usually occupied by guille-mots, the crevices by razorbills and perhaps puffins, while the kitti-wakes are seen on the little irregular scattered ledges, whether high up or low enough to be reached by the uptossed spray of the sea. But there is no hard and fast rule.   Guillemots or razorbills may be found nesting in just the places one would expect to find kittiwakes, and *vice versa*.   I have seen the kittiwakes on the same ledge with guillemots.   They have been observed to occupy, as a rule, the lowest positions on the cliff. This is the case at the Bass Rock, but at the Flamborough cliffs (Bempton) in Yorkshire, where I paid particular attention to the point, I did not find it so. The relative position of kittiwakes and guillemots on two of these cliff faces is shown by the following diagrams.

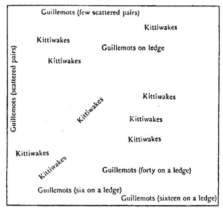

DIAGRAMS TO SHOW RELATIVE POSITIONS OF KITTIWAKES AND
GUILLEMOTS ON TWO CLIFF FACES AT BEMPTON, YORKS.

On both these cliffs the guillemots occupy the lowest and the highest places. In the case of Cliff A, this was simply due to the presence of large ledges near its base.[1]

A discovery forced upon my notice at these same cliffs, was that the kittiwakes nested only on their southerly or south-easterly faces : I was nowhere able to watch them without fronting a biting north-easterly wind. The guillemots and razorbills were quite indifferent as to aspect, a fact which made watching them in the early part of the year a matter of comparative comfort, as one could sit so as to be sheltered from wind and rain. On the Bass Rock, however, they nest on ledges facing all aspects, but chiefly south-east and north-west.

Laying begins toward the end of May, both sexes sharing in incubation, which lasts from three to four weeks. As is the case with many other species, there are usually a certain number of non-breeding adult pairs and single birds, which cannot apparently find room to nest, but exact information on this matter is still wanting. That these birds, or some of them, are willing to assume the duties of parenthood was shown by Faber, who removed a pair from a nest containing eggs, which were thereupon incubated by another pair, and the young hatched and reared. He found the same to be the case with puffins, guillemots, little-auks, and others (*Fratercula, Uria, Mergulus*).[2]

The young are fed by both parents on regurgitated food. The young bird shows its desire for food in a somewhat curious fashion ; it stands in its usual position, but with the head and neck contracted or drawn back between the shoulders. The head and beak are tilted slightly up, and bobbed up and down with a regular automatic action, that gives the bird somewhat the appearance of being a child's toy worked by internal mechanism. Usually the young kittiwake interrupts the movement from time to time to make a dab with

[1] Some years ago Mr. Jourdain noted that on the Bempton and Speeton Cliffs the kitti-wakes nested on the lowest ledges, none high up.
[2] Naumann, *Vögel Mitteleuropas*, xi. 293. For examples of the same habit among Passerine species, see above vol. i. 54, 55.

its beak at the beak of the parent. On one occasion I saw one adult and two young, all three side by side, and facing the rock face. Both the young were bobbing their heads, and the one nearest the parent bird turned at intervals to peck lightly at its beak. The second, however, kept on bobbing at the rock face, and continued this unprofitable exercise for some minutes before it turned to make a dab which did not even reach its mark. At the same time, on a lower ledge, another young bird kept on bobbing to the rock without once turning its head, while near by its parent stood preening itself. They were both at it some minutes later, one bobbing, the other preening, and I left them at it. The young will also bob in the same way when no other bird is near.

The actual feeding of the young by the parent I will illustrate by a fairly typical incident, noted at the time of its occurrence. A young bird was standing in front of its mother—it may have been the father—bobbing its head, occasionally making dabs at her beak. At length she opened her beak, and the young put its beak and the front part of its head inside and felt about. It seemed to get a small amount of something out of her throat and swallowed it down. After two or three more similar attempts, it pulled out what seemed like a large sand-eel, which I saw distinctly in its beak, and which it took some time to swallow down. It then continued to bob and dab.

The parent, as far as my observation extends, does not disgorge the food; it brings it up the gullet far enough to be reached by the young, and, frequently, it appears unable or unwilling to keep it in position, for the attempt is followed by visible gulping movements, which are similar to the movements of the young when actually swallowing. These swallowing movements resemble those already described provisionally as domestic amenities (p. 187). How or why the act of swallowing should have come to serve, as appears to be the case, the secondary or derived purpose of a gesture, and why this should have occurred rather in the case of the kittiwake than of other species, is difficult to divine.

Plate 108
Kittiwakes
By A. W. Seaby

Like other seabirds, kittiwakes go long distances for the food which they give to their young. It consists chiefly of small fish and crustaceans, which form also the main diet of the parents. The birds recorded by Collett and Nansen in the North Polar regions round Franz-Josef Land fed upon floating crustaceans, darting down upon them "with a dull splash against the surface of the water." Saxby found fresh-water weeds and a few beetles, probably picked up with the weeds, in the stomachs of birds he shot, but, this apart, the species appears to confine its diet to marine organisms. In this it differs from the species of *Larus*, which are omnivorous. It also differs from them in its method of catching fish. While they are content to sweep down to the water, and immerse only head and neck, it drops in head first, like a tern or gannet, from a height of twenty feet or so, cleaves the surface with a splash, and usually disappears completely from sight. It has been seen by Mr. H. Evans winging its way under the water.[1]

When the kittiwake pauses in its flight before dropping, it keeps its head windward. In consequence of this habit, flocks, when following a shoal of fish, work their way through it till they reach its windward side, and then make a wide flight back, and begin once more on the leeward margin.[2]

In August the young kittiwakes quit the cliffs, and are then to be seen in the neighbouring waters usually in flocks by themselves, with sometimes a few old birds among them. The gregarious instinct, therefore, asserts itself early. That it should at first take the form of association with birds of the same age is probably due to weakness of wing, the young birds not being able to follow the old long distances out to sea.

At this stage the young are seized and eaten by the larger Gulls, as already noted (p. 168), and the sight of one of these soft, dove-like creatures strangled and pick-axed by the strong hard beak of its

---

[1] J. H. Harvie-Brown and T. E. Buckley, *Fauna of Argyll and the Inner Hebrides*, p. 194.
[2] Saxby, *Birds of Shetland*, p. 330.

relentless murderer is among the most revolting and gruesome that Nature has to offer. The old birds are persecuted by the Skuas, and no doubt also at times fall victims to large Gulls and Hawks. Thousands used to be slaughtered for the decoration of ladies' hats, but fortunately the law stepped in to check this murderous form of feminine weakness, with the result that there are fewer dead kittiwakes in the shops and more alive on the sea.

Old and young quit their breeding-haunts towards the end of August, and are subsequently found all round our coasts and out to sea, sometimes in huge flocks. They are not, as far as I know, found inland except when storm-driven. Many are so worn out by heavy gales that they succumb, and their bodies are tossed ashore sometimes in hundreds—a feast for hungry Gulls and Crows.

Egg Plate F

1. Nightjar, 2 figures
2. Hoopoe, 1 figure
3. Pallas's sandgrouse, 2 figures
4. Puffin, 1 figure
5. Razorbill, 3 figures
6. Black-guillemot, 2 figures

CPSIA information can be obtained
at www.ICGtesting.com
Printed in the USA
BVHW03s1720260418
514512BV00014B/404/P